Field Programmable Gate Arrays (FPGAs) II

Edited by George Dekoulis

Published in London, United Kingdom

IntechOpen

Supporting open minds since 2005

Field Programmable Gate Arrays (FPGAs) II
http://dx.doi.org/10.5772/intechopen.78489
Edited by George Dekoulis

Contributors
J. Guadalupe Velasquez, Outmane Oubram, Luis Cisneros-Villalobos, Yongbo Liao, Mário Lopes Ferreira, João Canas Ferreira, Felipe A.P. de Figueiredo, Fabbryccio A.C.M. Cardoso, Surya Prasada Rao Borra, Rajesh K. Panakala, Pullakura Rajesh Kumar

First published in London, United Kingdom, 2020 by IntechOpen
IntechOpen is the global imprint of INTECHOPEN LIMITED, registered in England and Wales, registration number: 11086078, 7th floor, 10 Lower Thames Street, London, EC3R 6AF, United Kingdom
Printed in Croatia

British Library Cataloguing-in-Publication Data
A catalogue record for this book is available from the British Library

Additional hard and PDF copies can be obtained from orders@intechopen.com

Field Programmable Gate Arrays (FPGAs) II
Edited by George Dekoulis
p. cm.
Print ISBN 978-1-83881-056-6
Online ISBN 978-1-83881-057-3
eBook (PDF) ISBN 978-1-83881-058-0

For EU product safety concerns:
IN TECH d.o.o., Prolaz Marije Krucifikse Kozulić 3, 51000 Rijeka, Croatia, info@intechopen.com or visit our website at intechopen.com.

Meet the editor

Prof. George Dekoulis received his PhD in Space Engineering and Communications from Lancaster University, UK, in 2007. He was awarded a 1st Class BEng (Hons) degree in Communications Engineering from De Montfort University, UK, in 2001. He has received several awards from STFC, UK and EPSRC, UK, and the "IET Hudswell International Research Scholarship". He is currently a professor at the Aerospace Engineering Institute (AEI), Cyprus. He is founder of the IEEE Aerospace and Electronic Systems Society (AESS) – Cyprus and was the General Chair of IEEE Aerospace Engineering Innovations 2019 (IEEE AEI 2019) Symposium, 20-23 April 2019, Limassol, Cyprus. He has previously worked as a professor in aerospace engineering at various departments, such as space and planetary physics, aeronautical and space engineering, professional flight, robotics/mechatronics and mechanical engineering, computer science and engineering and electrical and electronics engineering. His research is focused on the design of reconfigurable aerospace engineering systems.

Contents

Preface

This Edited Volume is a collection of reviewed and relevant research chapters, concerning the developments within the Field Programmable Gate Arrays (FPGAs) II area of study. The book includes scholarly contributions by various authors and edited by a group of experts pertinent to Computer and Information Science. Each contribution comes as a separate chapter complete in itself but directly related to the book's topics and objectives.

The book is divided in one section which includes chapters dealing with the topics: Real-Time FPGA-Based Systems to Remote Monitoring, Real-Time Echo State Network Based on FPGA and Its Applications, Flexible Baseband Modulator Architecture for Multi-Waveform 5G Communications, An Efficient FPGA-Based Frequency Shifter for LTE/LTE-A Systems, and VLSI Implementation of Medical Image Fusion Using DWT-PCA Algorithms.

The target audience comprises scholars and specialists in the field.

George Dekoulis
Aerospace Engineering Institute,
Cyprus

Chapter 1

Real-Time FPGA-Based Systems to Remote Monitoring

J. Guadalupe Velásquez-Aguilar, Outmane Oubram and Luis Cisneros-Villalobos

Abstract

Some industrial and laboratory applications such as control, monitoring, test and measurements, and automation require real-time systems for their development. Embedded systems for acquisition and processing often require the participation of the embedded operating system and therefore are necessary techniques that can accelerate software execution. The latest field-programmable gate arrays' (FPGA) technology has blurred the distinction between hardware and software with embedded processors that allow the development of Systems-on-a-Chip (SoC) running on operating systems. The widespread adoption of wireless technologies such as Bluetooth, ZigBee, and Wi-Fi in the last years has facilitated the use of these technologies to the development of real-time monitoring applications that combined with FPGA devices which has the advantages of low cost, flexibility, and scalability as compared with other commercial systems.

Keywords: real-time monitoring, wireless FPGA-based controllers

1. Introduction

In many of the industrial and laboratory systems, especially in control and monitoring tasks, hardware is used in a loop (**Figure 1**).

In the diagram shown above, information about the physical environment is obtained through sensors that respond to a physical stimulus (light, heat, pressure, magnetism, acceleration, stress) and that are designed so that the information acquired is transformed into an electrical signal proportional to the changes. Frequently, the electrical signal obtained from the sensor has noise or interference, so signal conditioning is necessary, which is achieved through some processing operations such as amplification, linearization, compensation, and filtering. Analog-to-digital converters (ADC) are used to sample and hold charge, thereby converting the analog circuit current/voltage into a digital value. Without encoding, sensors are useful in analog control systems, but for the use in digital control and monitoring systems, encoding is critical. Real-time embedded systems therefore require digital encoding of all sensor inputs, with the exception of subsystems, which are all analog [1].

One of the main tasks of embedded systems is the processing and interpretation of information that arrives from the outside. An embedded system is a combination of hardware and software that is specifically designed for a particular function. In most cases, an embedded system is used to replace an application specific

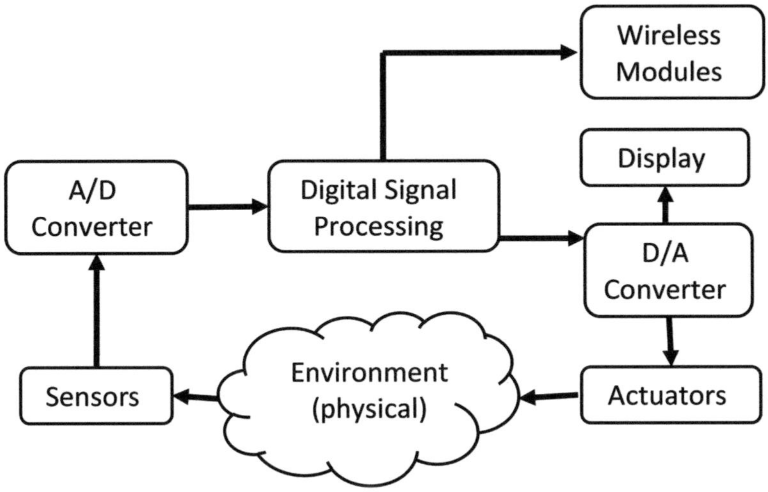

Figure 1.
Real-time processing system hardware in the loop.

electronics in consumer products. By doing so, most of the systems functionality is encapsulated in the firmware that runs the system, and it is possible to change and upgrade the system by changing the firmware, while keeping the hardware same [2]. When embedded systems are board-based, it is fairly straightforward to select the proper components, integrate them with software, and ship the product.

In the mid-1990s, the development of embedded systems evolved with the concept of ASIC technology, changing the philosophy of systems based on a chip-set to a concept System-on-a-Chip (SoC) based on embedded cores. The term SoC defines an integrated circuit (IC) designed by joining multiple independent VLSI models to provide full functionality for an application. Each model is predesigned with complex functions known as cores that serve to a variety of applications. Cores can use a library of components designed by intellectual property (IP) companies or by self in house. The chip used for the system may contain combinations of cores that are generally available in the form of a synthesizable high-level description language (HDL), as Verilog/VHDL, or optimized transistor-level design. Some examples of core-based SoC include high-end microprocessors, GPS positioning for autonomous vehicles, smartphone, and even PC-on-a-Chip [3].

Nowadays, embedded systems are made on SoC. The SoC can include several heterogeneous subsystems, including specific hardware components and sophisticated interconnects (**Figure 2**).

Often in systems used in industrial and laboratory applications for control, monitoring, testing and measurement, and automation, data acquisition (DAQ) subsystem is the first stage. The main purpose of DAQ is to measure physical phenomena, converting the analog signal into a digital signal, and then send or save the data collected for further analysis. An important point to consider is the problems of output conversion into a digital format, as well as to high accuracy and speed conversion methods used. In addition, if the application requires simultaneously capturing several signals, the DAQ must be of the multichannel type and will need a central processor, which will control the channeling and organization of data acquisition for further displays or its use in control systems. The methods to be used in multichannel data acquisition depend on the control and measurement tasks and directly influence the structure and functionalities of the DAQ. In a modern system, the measurement and control sensors can be set up in different ways; the most used are:

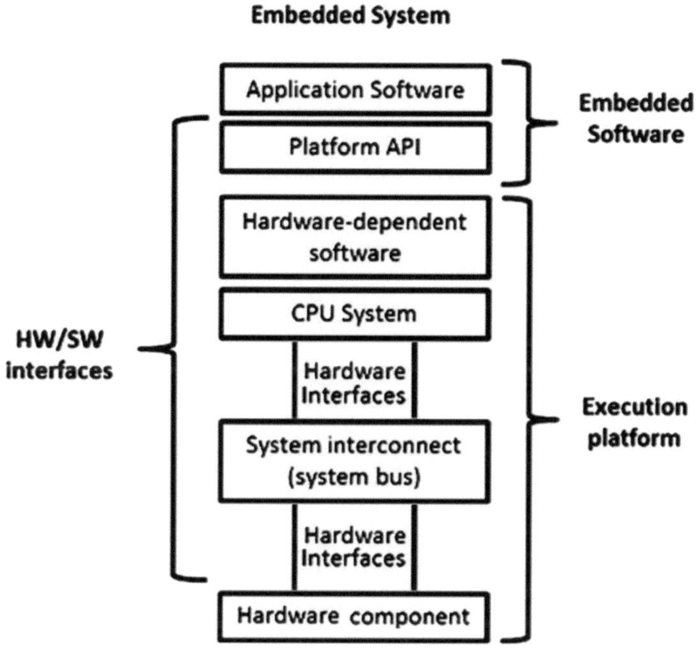

Figure 2.
Embedded system architecture. In addition to hardware, a SoC includes classic application software- and hardware-dependent software that must be co-designed with hardware interfaces. The API hides hardware details such as interrupt controllers or memory and I/O systems [4].

- Methods that use time-division channeling, which perform sensor multiplexing, that is, the time is shared by each sensor in the data acquisition.

- Methods using space-division channeling, based on simultaneous data acquisition from all the sensors.

In both cases, access to information at any time depends on the control and measurement tasks used [5].

Commercial DAQ cards are differentiated by their viabilities such as sampling frequency, scale of acquired signal, power, and requirements but are generally high in cost, and they need a PC at the collection site. Embedded systems to data acquisition often require the participation of the embedded operating system. The modern on-board FPGA can not only overcome the deficiency of the microcontroller unit (MCU) or the digital signal processor (DSP) and meet the requirements of system for real-time and synchronization but also for embedded applications using SoC FPGA platforms with the high level coordination, versatility, and full-stacked operative system [6].

2. Wireless communications standard protocols

At present, the most used standard protocols in communication in wireless sensor networks (WSN) are IEEE 802.15.1 Bluetooth, IEEE 802.15.4, IEEE 802.15.4/a ZigBee, and IEEE 802.11 Wi-Fi. The following describes these protocols:

2.1 Bluetooth technology

IEEE 802.15.1 protocol is an economical and secure wireless communication standard, used to exchange information between devices through a short-range radio

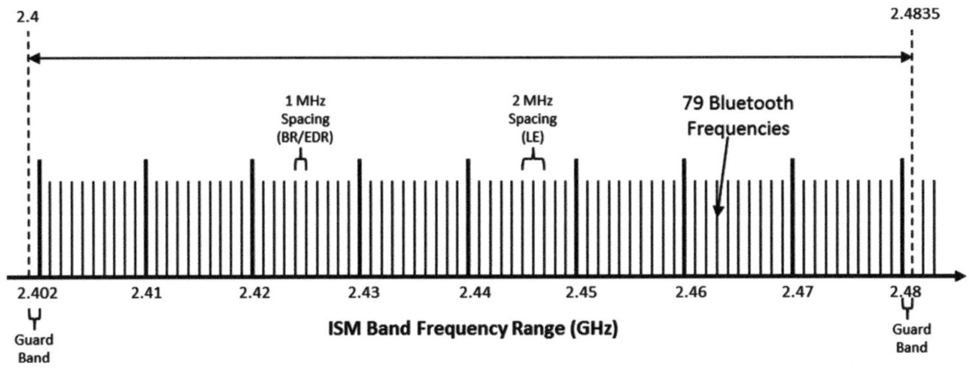

Figure 3.
Bluetooth frequency bands and RF channels. Each RF channel is ordered in channel number n as follows: $f = 2402 + n$ *MHz, where* $n = 0, \ldots, 78$ *(BR/EDR) and* $f = 2402 + n^{*}2$ *MHz, with* $n = 0, \ldots, 39$ *(LE).*

frequency; it was invented in 1994 by a group of engineers of the Ericsson Company. The original idea of Bluetooth was to eliminate the need for a cable connection between devices by connecting them over short distances (up to 100 m). Bluetooth operates with industrial, scientific, and medical frequencies (ISM), from 2.4 to 2.4835 GHz starting at 2.402 GHz. Bluetooth devices can be configured to operate in two ways:

1. Basic and Enhanced Data Rates (BR/EDR) transmissions, where 79 radio frequency (RF) channels with 1 MHz spacing are used. This configuration uses frequency-hopping spread spectrum (FHSS) scheme, at a nominal rate of 1600 hop per second.

2. Low Energy (LE) mode, where only 40 RF channels with 2 MHz spacing are available and adaptive frequency hopping (AFH) is used (**Figure 3**) [7, 8].

Since its appearance, Bluetooth protocol has continuously evolved, so there are several versions that are differentiated with a number. Bluetooth versions 1.0–3.0 are known as Bluetooth Classic category and originally supported a maximum data rate of 721 kbps. This is referred to as Basic Rate (BR). The Bluetooth 2.0 EDR specification added support for data rates up to 2.1 Mbps. This is referred to as Enhanced Data Rate (EDR). The Bluetooth 3.0 High Speed (HS) specification enhanced it even further to 24 Mbps. Bluetooth Low Energy (BLE) is a new category that include versions 4.0 and 5.0. Geared toward applications requiring low power consumption, BLE returns to a lower data throughput of 1 Mbps using the GFSK modulation scheme. The Bluetooth 4.0 specification did not add any additional data rates; it only reduced the current consumption to enable low-energy devices. In Bluetooth 5.0, in addition to low power consumption, four different data rates are offered to accommodate a variety of transmission ranges: 2 Mbps, 1 Mbps, 500 kbps, and 125 kbps. The lower data rate of 125 kbps was added to compensate for the increase in transmission range [9].

Bluetooth module generally consists of four components: radio transceiver, baseband/link controller, link manager, and a host controller interface (HCI) [8]. HCI is the interface to access the Bluetooth module setup from the host. Bluetooth communication is based on the following two network topologies:

1. Piconet: It consists of one master and up to seven slaves (**Figure 4a**).

2. Scatternet (combination or two or more piconets) (**Figure 4b**): It is formed when two or more piconets come together by sharing a device. Scatternets help

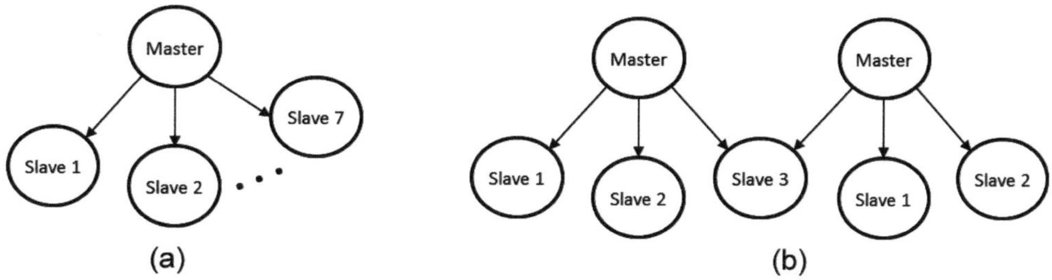

Figure 4.
Bluetooth network topologies. (a) Piconet. (b) Scatternet.

to extend the number of Bluetooth devices that can communicate with each other. They allow more than seven devices to communicate with each other [10].

2.2 ZigBee technology

ZigBee, also known as IEEE 802.15.4, was initially conceived in 1998, standardized in 2003, and finally revised in 2006; it is a low power standard for short-range communications between wireless devices. ZigBee is classified as a wireless personal area network (WPAN). ZigBee devices operate in one of three bands: 868 MHz (Europe), 915 MHz (North America), and 2.4 GHz (worldwide). The 2.4 GHz band is the most used by the ZigBee transceivers and uses offset quadrature phase-shift keying (OQPSK) modulation stream. This type of modulation, which is a derivation of traditional QPSK, is used for requiring less transmission power and achieving the same or better performance than similar ones. OQPSK modulation combined with the use of a 5 MHz channel bandwidth allows devices to reach a data rate of up to 250 kbits/s efficiently [11]. The IEEE 802.15.4 has three different operation modes (**Figure 5**):

1. Personal area network coordinator (ZigBee coordinator, ZC): It is the principal controller of the PAN. This device identifies the network, and in it the configurations that allow other devices to be associated are made. ZC function is to act as ZigBee Router (ZR) once the network is formed. ZC is a full-functional device (FFD) that implements the full protocol stack; it can operate with or without beacon mode. The beacon mode of operation is used when data packets must be sent within an allowable delay, such as in monitoring and control applications. The beaconless mode is suitable for applications where data is only sent when an event occurs, that is, there is no continuity in sending information such as motion detection. In a cluster-tree network, all ZRs will receive beacons from their parents and send their own beacons to synchronize the nodes that belong to their clusters.

2. Local Coordinator (ZigBee Router, ZR): This device must be associated with a ZC or with another ZR previously associated with a network, because it does not create its own network. ZR is a full-functional device (FFD) that implements the full protocol stack. This device participates in multi-hop routing of message in mesh and cluster-tree networks (in the latter case they are also called cluster heads (CHs)). ZR provides synchronization services through beacon transmission.

3. End device (ZigBee end device, ZED): It is a device that does not implement the previous functionalities and should associate with a ZC or ZR before interacting with the network. In ZigBee, it is just a sensor/actuator node; it can be a reduced function device (RFD) that implements a reduced subset of the protocol stack [12].

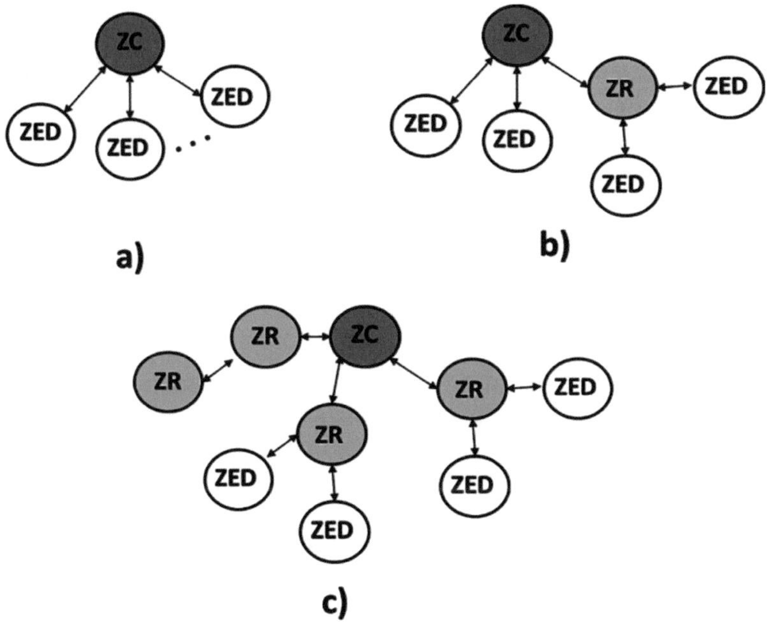

Figure 5.
ZigBee network topologies. (a) Star topology contains a unique node that operates as ZC, which establishes the PAN identifier. The identifier should not be used by any other ZigBee network in the vicinity. Also in the star topology, the communication is centralized, so each device (FFD or RFD) joining the network and willing to communicate with other devices must send its data to the ZC, which sends it to the adequate destination. (b) Mesh topology includes a ZC that identifies the entire network. Communication in this topology is decentralized, so each node can communicate directly with any other node within its radio. (c) In cluster tree topology, there is a single routing path between any pair of nodes, and there is a distributed synchronization mechanism (IEEE 802.15.4 beacon-enabled mode). There is only one ZC that identifies the entire network and one ZR per cluster. Any of the FFDs can act as a ZR that provides synchronization services to other devices and ZRs [13].

It is important to consider some operational considerations that may be presented by topologies for traditional wireless sensor networks (WSN). If you choose to use the star topology, you should keep in mind (a) that the sensor node selected as ZC will quickly consume its battery and (b) that the coverage of an IEEE 802.15.4/ZigBee cluster is very limited when addressing a large-scale WSN, leading to a scalability problem. On the other hand, the mesh topology enables enhanced networking, but it induces additional complexity to provide end-to-end connectivity between all nodes in the network. Therefore, unlike the star topology, the mesh topology can be more energy efficient, since the communication process does not depend on a particular node [14].

2.3 Wi-fi technology

Wi-Fi is the name given by the Wi-Fi Alliance [15] to the IEEE 802.11 suite of standards. 802.11 defined the initial standard for wireless local area networks (WLANs).

The evolution of Wi-Fi technology has focused on increasing speed, lower latency, and better user experiences in a multitude of environments and with a variety of device types. Wi-Fi Alliance has introduced generational names to devices and product descriptions. The latest generation of Wi-Fi devices, based on the 802.11ax standard, is known as Wi-Fi devices 6. If the device contains 802.11 ac, 5 GHz technology is known as Wi-Fi 5, or if the device uses technology 802.11n, 2.4 GHz is known as Wi-Fi 4 [16]. Generations of Wi-Fi prior to Wi-Fi 4 will not be assigned names. Most of devices available in the market today are identified as Wi-Fi 5.

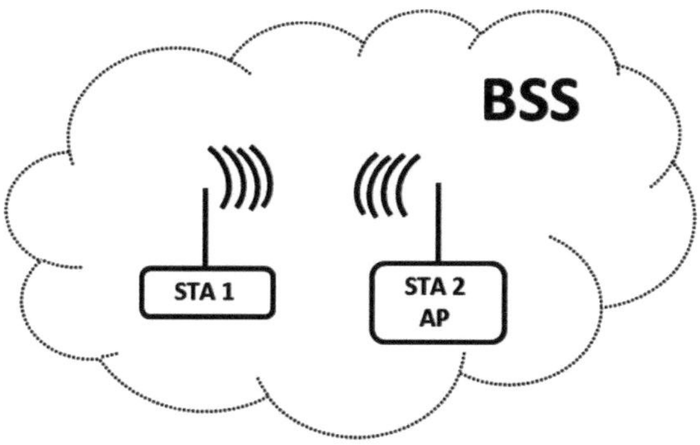

Figure 6.
BSS controlled by a single coordination function (CF). The CF determines when a STA transmits and when it receives.

Wi-Fi is a physical layer/link interface, as is Ethernet. A wireless station (STA) can be a personal computer (PC), a laptop, a personal digital assistant (PDA), or phone. When two or more STAs are connected wirelessly, they form a Basic Service Set (BSS) (**Figure 6**). This is the basic component of a Wi-Fi network [17].

Wi-Fi has two different operation modes: infrastructure mode and ad hoc mode. Each one uses the BSS, but they yield different network topologies.

1. Ad hoc mode: Wireless stations communicate directly with one another, with a peer-to-peer network model. A BSS operating in ad hoc mode is isolated, that is, there is no connection to other Wi-Fi or wired LAN networks. The utility of this network is in situations that demand a quick setup in places where there is no network infrastructure.

2. Infrastructure mode: This mode requires the BSS to contain a wireless access point (AP). An AP is an STA with additional functionality that allows extending access to wired networks for clients of a wireless network. Any wireless device that tries to join the BSS must first be associated with the AP. A distribution system (DS) is generated when an AP provides access to its associated STAs. The DS can allow communication between APs as shown in **Figure 7**.

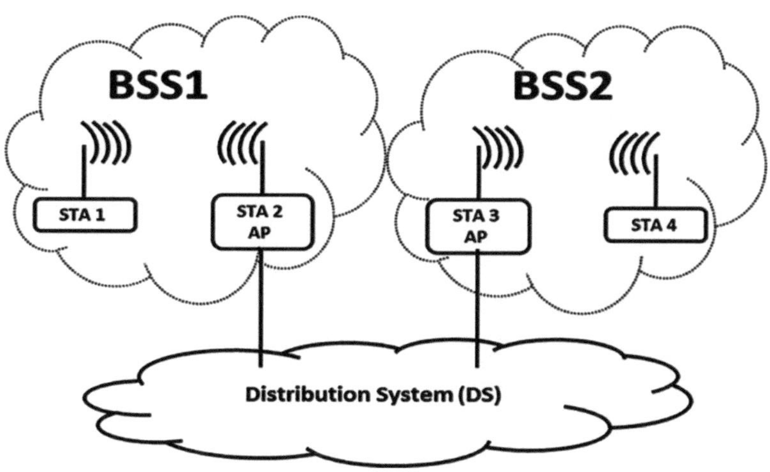

Figure 7.
All wireless communication to or from an associated STA goes through an AP. This type of setup is similar to the "star topology" used in wired networks.

2.3.1 Services specified by IEEE 802.11

The IEEE 802.11 standard does not define any specific implementations. Instead, nine services are specified that all implementations must provide; these are:

2.3.1.1 Station services (SS)

Authentication - The STA must identify itself to the AP before it can access network services.

De-authentication - This service voids an existing authentication.

Privacy - An STA must be able to encrypt the frame to protect the message content to be transmitted, so that only the recipient can read it.

MAC service data unit (MSDU) delivery - An MSDU is a data frame that must be transmitted to the proper destination.

2.3.1.2 Distribution system services (DSS)

A STA that functions as an AP must implement the following services:

Association - This service establishes an AP/STA mapping after mutually agreeable authentication has taken place between the two wireless stations. A STA can only associate with one AP at a time.

Re-association - This service allows you to change the current association from one AP to another AP.

Disassociation - This service voids a current association.

Distribution - This service handles delivery of MSDUs within the distribution system.

Integration - This service is the bridge function between wireless and wired networks. MSDU handles the delivery of between the distribution system and a wired LAN [17].

3. Hardware description

The elements used for the realization of the proposed system are shown in **Figure 8**. The platform is composed of four components: the FPGA board that includes A/D converter and three wireless interface Bluetooth, XBee (ZigBee protocol-based), and Wi-Fi module. The wireless modules provide the FPGA device the capacity to communicate with other system or the Internet.

Figure 8.
Hardware components used for the real-time monitoring system.

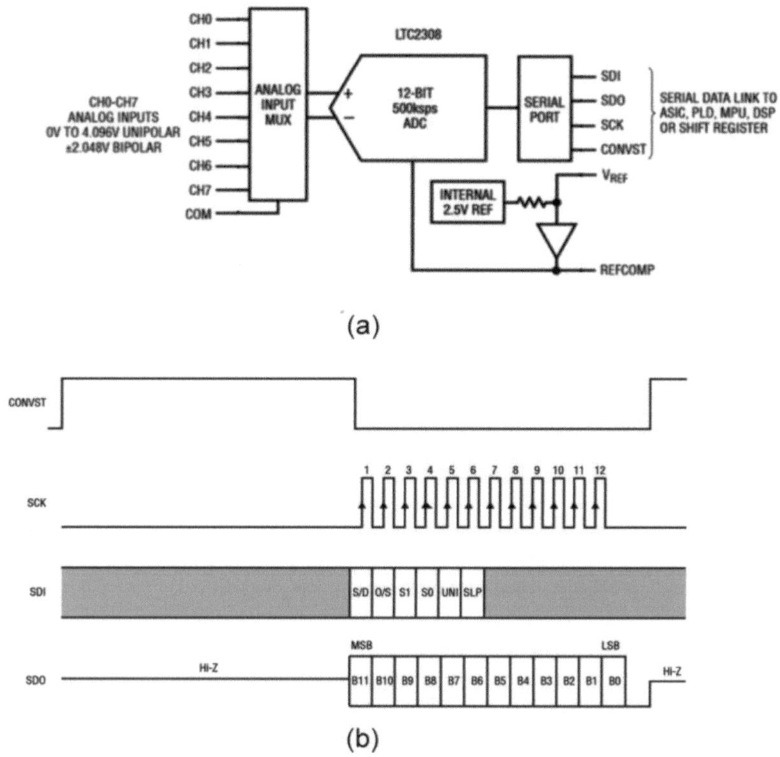

Figure 9.
(a) Block diagram LTC2308 device. Eight analog input and operation modes can be programmed by a 6-bit DIN word through SDI terminal. (b) Timing with a long pulse. The configuration signals are S/D can be single-ended/differential-bit; O/S can be odd/sing-bit; S1 and S0 addressing select bit; UNI can be unipolar/bipolar and SLP active sleep mode [19].

3.1 Analog/digital converter

The A/D converter chip used is the integrated circuit (IC) LTC2308, Linear Technology, whose characteristics are low noise and power consumption, up to 500 Kbps, 8-channel, 12-bit, and SPI/MICROWIRE compatible serial interface. The internal conversion clock allows the external serial output data clock (SCK) to operate at any frequency up to 40 MHz [18]. **Figure 9** shows the block diagram of ADC.

3.2 FPGA device

Cyclone V FPGA. The Intel FPGA Cyclone V SE 5CSEMA4U23C6N device has Dual ARM Cortex-A9 MPCore with Core Sight System on Chip (SoC), integrated circuit Cyclone V SE FPGA, with 40 K logic elements, maximum CPU clock frequency 925 MHz, 224 18×19 multipliers, and 5761 kb embedded memory (**Figure 10**).

3.3 Wireless modules

3.3.1 XBee Pro S1

The Digi XBee series modules implement the IEEE 802.15.4 radio and ZigBee networking protocol for its physical layer and MAC. Outdoor transmission distances to 0–90 meters depending on power output and environmental characteristics. XBee devices work in ISM 2.4 GHz frequency bands having a serial interface data

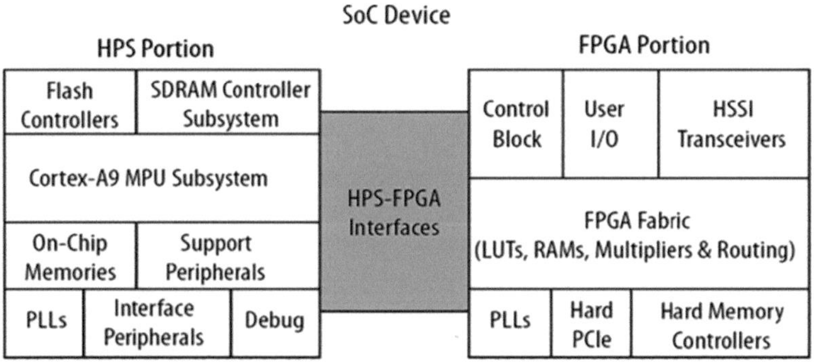

Figure 10.
Cyclone V SoC device block diagram is composed of two distinct portions: A dual-core ARM cortex-A9 hard processor system (HPS) and an FPGA. The cortex-A9 processor has two 32-bit CPUs and associated subsystems on the Intel Cyclone V SoC chip, where hardware circuits can be implemented, which reduce the size of the board and increase the performance of the developed system [20].

rate from 1200 bps to 250Kbps. The following are the supported network topologies: point-to-point, point-to-multipoint, and peer-to-peer.

3.3.2 HC-06 Bluetooth 2.0 EDR module

This module is a serial interface converter to Bluetooth adapter. HC-06 has a 2.4GHz digital wireless transceiver, low power consumption, an EDR module, the change range of modulation depth: 2Mbps–3Mbps, and standard HCI Port (UART or USB), and it can work at the low voltage (3.1–4.2 V). The module can be set by AT commands and have two modes, master and slave, but the mode cannot be switched during the process of communication. Serial baud rate is 1200–1,382,400 bps [21].

3.3.3 ESP8266 Wi-Fi module.

This module implements TCP/IP and full 802.11 b/g/n (support 2.4 GHz, up to 72.2 Mbps) WLAN MAC protocol. It can perform either as a stand-alone application or as the slave to a host MCU, so it supports Basic Service Set (BSS) STA and SoftAP operations under the distributed control function (DCF). ESP8266 includes a CPU Tensilica L106 32-bit processor, and it has peripheral interfaces: UART, SDIO, SPI, I2C, I2S, and IR. Power management is handled with minimum host interaction to minimize active duty period. ESP8266EX can be applied to any microcontroller design as a Wi-Fi adaptor through SPI/SDIO or UART interfaces [22].

4. System architectures

The design of an FPGA-based remote monitoring system architectures is show in **Figure 11**. The resultant design is implemented in VHDL and block diagrams; it is validated in co-simulation environment, and finally, it is tested in a real-time application to monitoring an electric signal.

There are three important features to consider before starting the development system: first, the nature of the feedback signal. If the sensor which measures the variable to be monitored has an analog nature, it is necessary to use an analog-to-digital converter (ADC) which has an output with a fixed bit width. Second: in order to avoid performing arithmetic operations between signals of different bit

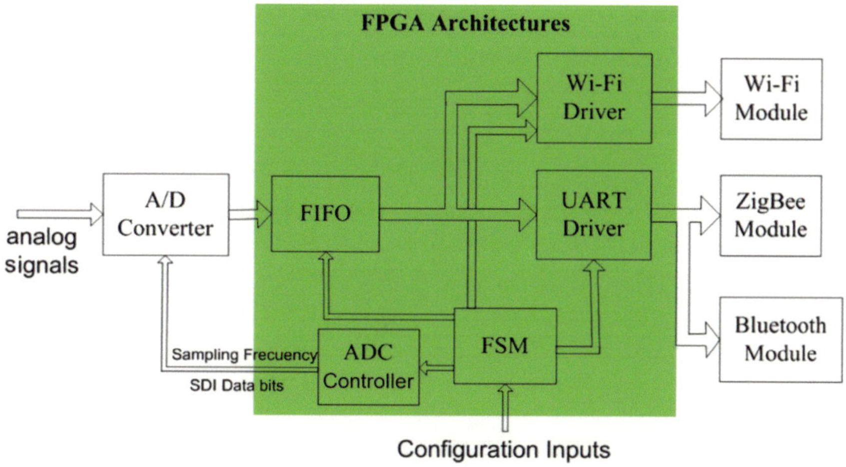

Figure 11.
Block diagram of architectures implemented on FPGA. This module comprises five blocks: ADC controller, FIFO memory, Wi-fi, UART drivers, and a finite state machine (FSM).

width, it is strongly recommended that the operations have the same bit width as the measured variable. Finally, the system output must be congruent with the bus width wireless interface.

4.1 A/D converter controller

The ADC LTC2308 operates on a 12-cycle operational frame, as shown in **Figure 9b**. ADC has four wires to control and communicate with the FPGA: SCLK, CS, DIN, and DOUT. The SCLK and CS signals are used to control the ADC. SCLK is the signal clock for the ADC. The CS signal serves as chip select for the ADC chip. The DIN and DOUT wires are used for transferring addresses and data between the two chips (ADC and FPGA). The FPGA uses the DIN connection to provide the address (3 bits in length) of the next channel requested for conversion. The DOUT connection is used by the ADC to send the digital value (12 bits long) of the converted signal to the FPGA. Both DIN and DOUT are sent in a serial manner at a rate of 1 bit per SCLK cycle [23].

In the case of our working example, SPI controller was developed to control the conversion process. A long CONVST pulse is used. **Figure 9b** shows time diagram to programming ADC. According to the diagram, "the conversions are initiated by a rising edge on the CONVST input. Once a conversion cycle has begun, it cannot be restarted. Between conversions, a 6-bit input word (DIN) at the SDI input configures the MUX and programs various modes of operation. As the DIN bits are shifted in, data from the previous conversion is shifted out on SDO. After the 6 bits of the DIN word have been shifted in, the ADC begins acquiring the analog input in preparation for the next conversion as the rest of the data is shifted out" [19]. **Figure 12** shows the block diagram architecture corresponding to SPI controller.

4.2 FIFO architecture

A dual-clock First-In First-Out (FIFO) buffer was used to cross data between the two different clock domains: sampling frequency A/D converter (from 1 to 25 MHz) and transmission rate (from 9600 to 921,600 bps), **Figure 13**. In the systems' clock frequency domain, the serialized outputs are continuously stored in 12 bits shift

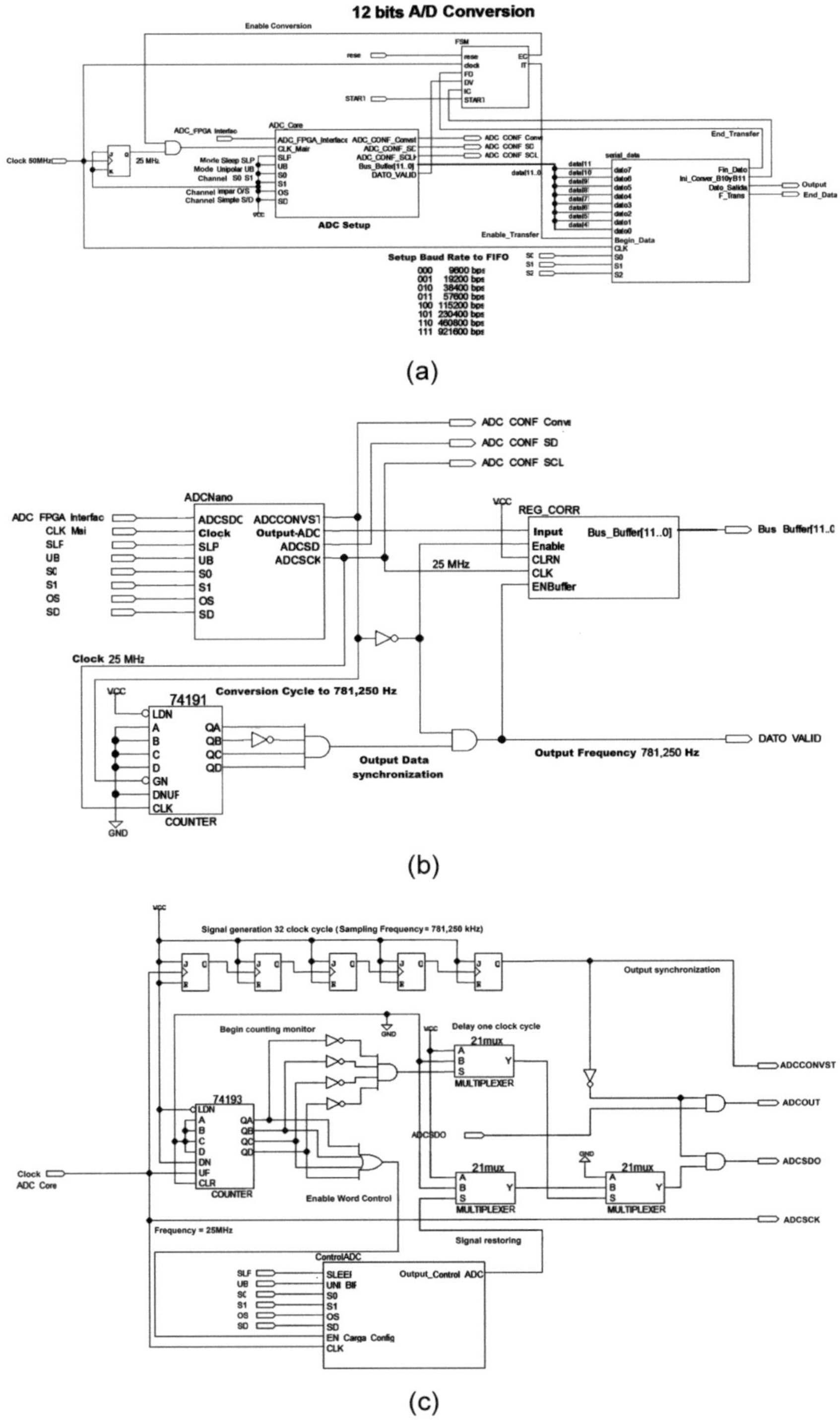

Figure 12.
SPI controller architecture. (a) 12 bits a/D conversion general architecture. (b) ADC_Core architecture. (c) ADC_Nano architecture generates signal control to ADC. The 4-bit counter counts 16 cycles in high for the acquisition of the signal and 16 cycles in low for the sending of the 12 output bits parallel to the configuration instruction for the next sample. The control ADC architecture is based on shift register.

register, before they will be sent to FIFO buffer. The finite state machine (FSM) FIFO, in the system controller, wait until collected data of the last active channel will be sent through wireless module, before starting a new acquisition.

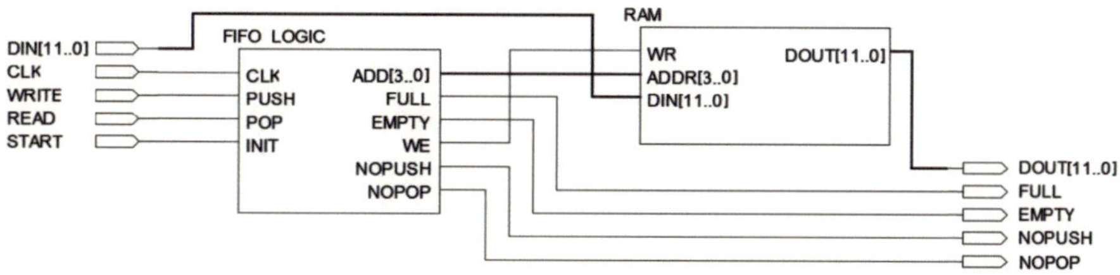

Figure 13.
Dual-clock FIFO architecture. Two counters are used to addressing the data to read and write operations. RAM of 12-bit and 16 words is used to store data.

The following source code corresponds to the FIFO_LOGIC and RAM entities of the design.

Code 1. FIFO_LOGIC.vhd [24].

```
LIBRARY IEEE;
USE IEEE.STD_LOGIC_1164.ALL;
USE IEEE.STD_LOGIC_UNSIGNED.ALL;

ENTITY FIFO_LOGIC IS
    PORT (
        CLK, PUSH, POP, INIT:               IN      STD_LOGIC;
        ADD:                                OUT     STD_LOGIC_VECTOR(3 DOWNTO 0);
        FULL, EMPTY, WE, NOPUSH, NOPOP: BUFFER  STD_LOGIC);
END ENTITY FIFO_LOGIC;

ARCHITECTURE BEHAVIOR OF FIFO_LOGIC IS
    SIGNAL WPTR, RPTR:  STD_LOGIC_VECTOR (3 DOWNTO 0);
    SIGNAL LASTOP:      STD_LOGIC;
BEGIN
    SEQUENTIAL: PROCESS (CLK)
    BEGIN
        IF (CLK'EVENT AND CLK = '1') THEN
            IF (INIT = '1') THEN
                WPTR <= (OTHERS => '0');
                RPTR <= (OTHERS => '0');
                LASTOP <= '0';
            ELSIF (POP = '1' AND EMPTY = '0') THEN              -- POP
                RPTR <= RPTR + 1;
                LASTOP <= '0';
                ELSIF (PUSH = '1' AND FULL = '0') THEN          -- PUSH
                    WPTR <= WPTR + 1;
                    LASTOP <= '1';
                END IF;
        END IF;
    END PROCESS SEQUENTIAL;

    COMBINATIONAL: PROCESS (PUSH, POP, WPTR, RPTR, LASTOP, FULL, EMPTY)
    BEGIN
        IF (RPTR = WPTR) THEN
            IF (LASTOP = '1') THEN
                FULL <= '1';
                EMPTY <= '0';
            ELSE
                FULL <= '0';
                EMPTY <= '1';
            END IF;
        ELSE
            FULL <= '0';
            EMPTY <= '0';
        END IF;

        IF (POP = '0' AND PUSH = '0') THEN          -- NO OPERATION --
            ADD <= RPTR;
            WE <= '0';
            NOPUSH <= '0';
            NOPOP <= '0';
        ELSIF (POP = '0' AND PUSH = '1') THEN       -- PUSH ONLY --
            ADD <= WPTR;
            NOPOP <= '0';
            IF (FULL = '0') THEN                    -- VALID WRITE CONDITION --
                WE <= '1';
                NOPUSH <= '0';
            ELSE                                    -- NO WRITE CONDITION --
                WE <= '0';
                NOPUSH <= '1';
            END IF;
            ELSIF (POP = '1' AND PUSH = '0') THEN -- POP ONLY --
                ADD <= RPTR;
                NOPUSH <= '0';
                WE <= '0';
                IF (EMPTY = '0') THEN               -- VALID READ CONDITION --
                    NOPOP <= '0';
                ELSE
                    NOPOP <= '1';                   -- NO READ CONDITION --
                END IF;
            ELSE                                    -- PUSH AND POP AT SAME TIME -
                IF (EMPTY = '0') THEN               -- VALID POP --
                    ADD <= RPTR;
                    WE <= '0';
                    NOPUSH <= '1';
                    NOPOP <= '0';
                ELSE
                    ADD <= WPTR;
                    WE <= '1';
                    NOPUSH <= '0';
                    NOPOP <= '1';
                END IF;
            END IF;
    END PROCESS COMBINATIONAL;
```

Code 2. RAM_16.vhd.

```
LIBRARY IEEE;
USE IEEE.STD_LOGIC_1164.ALL;
USE IEEE.STD_LOGIC_UNSIGNED.ALL;

ENTITY RAM_16 IS
    PORT
    (
        WR:      IN STD_LOGIC;
        ADDR:    IN STD_LOGIC_VECTOR (3 DOWNTO 0)
        DIN:     IN STD_LOGIC_VECTOR (11 DOWNTO 0
        DOUT:    OUT STD_LOGIC_VECTOR (11 DOWNTO
    );
END ENTITY RAM_16;

ARCHITECTURE BEHAVIOR OF RAM_16 IS
SUBTYPE WORD IS STD_LOGIC_VECTOR (11 DOWNTO 0);
TYPE MEMORY IS ARRAY (0 TO 15) OF WORD;
SIGNAL RAM16: MEMORY;
BEGIN
    PROCESS (WR, DIN, ADDR)
    VARIABLE RAM_ADDR_IN: INTEGER RANGE 0 TO 15;
    BEGIN
        RAM_ADDR_IN := CONV_INTEGER (ADDR);
        IF (WR = '1') THEN
            RAM16 (RAM_ADDR_IN) <= DIN;
        END IF;
        DOUT <= RAM16(RAM_ADDR_IN);
    END PROCESS;
END ARCHITECTURE BEHAVIOR;
```

4.3 UART driver

Serial communications depend on the two UART devices (the FPGA architecture and the wireless module) to be configured with compatible settings: baud rate, parity, control (start and stop bits), and data bits (**Figure 14**).

In this system, a general port input/output (GPIO) is used to send serial data. Subsystem architecture (**Figure 15**) is used to set the baud rate in the output. UART interface will read out the data when it is filled in the FIFO and send to the host

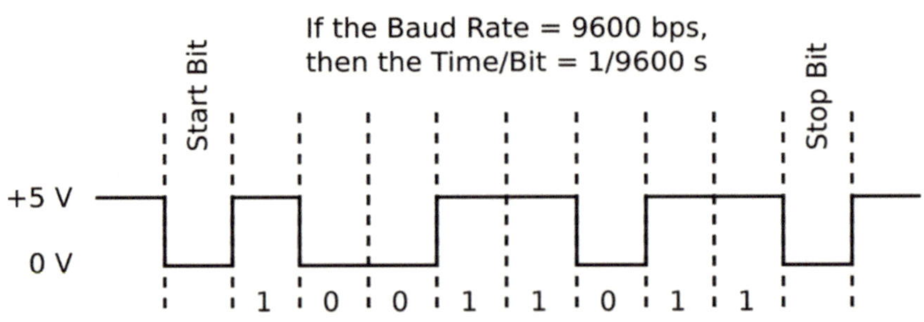

Figure 14.
UART data packet has data format structure: Data bits, parity, and stop bits. In the graph, the data 0x9B (decimal number "155," ASCII character "ø") is transmitted through the wireless module with format: 8-N-1 [25].

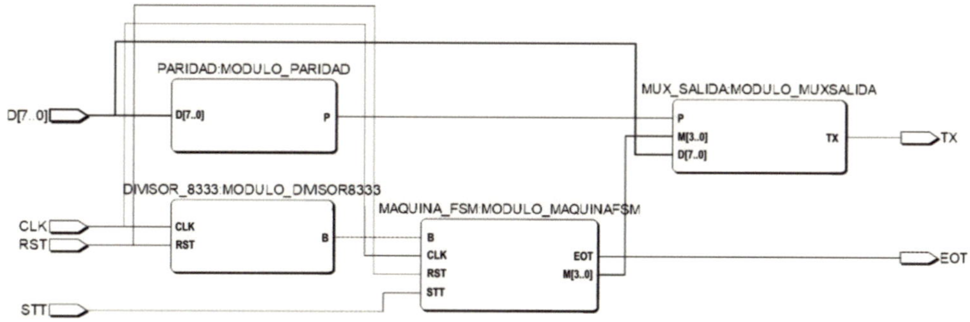

Figure 15.
UART driver diagram. Serial transmission uses baud rate module (DIVISOR_8333). MAQUINAFSM together with MUXSALIDA sends data from FIFO to serial data in the transmission format. The parity is verified with PARIDAD.

through the wireless link (Bluetooth or XBee modules), and finally the data can be displayed in the host with software application.

Code 3. DIVISOR_8333.vhd.

```
USE IEEE.STD_LOGIC_1164.ALL;
USE IEEE.STD_LOGIC_ARITH.ALL;
USE IEEE.STD_LOGIC_UNSIGNED.ALL;

ENTITY DIVISOR_8333 IS
  PORT(
  RST  : IN  STD_LOGIC;
  CLK  : IN  STD_LOGIC;
  B    : OUT STD_LOGIC
  );
END DIVISOR_8333;

ARCHITECTURE BEHAVIOR OF DIVISOR_8333 IS
SIGNAL QN, QP, A : STD_LOGIC_VECTOR(13 DOWNTO 0);
BEGIN
  A <= (OTHERS => '0');
  COMBINACIONAL:PROCESS(QP,A)
  BEGIN
    IF(QP = A) THEN
      B  <= '1';
      QN <= "01010001010111";
    ELSE
      B  <= '0';
      QN <= QP - 1;
    END IF;
  END PROCESS COMBINACIONAL;

  SECUENCIAL:PROCESS(RST,CLK)
  BEGIN
    IF(RST = '1')THEN
      QP <= (OTHERS => '0');
    ELSIF(CLK'EVENT AND CLK = '1')THEN
      QP <= QN;
    END IF;
  END PROCESS SECUENCIAL;
END BEHAVIOR;
```

Code 4. MUX_SALIDA.vhd.

```
LIBRARY IEEE;
USE IEEE.STD_LOGIC_1164.ALL;

ENTITY MUX_SALIDA IS
  PORT(
  M  : IN  STD_LOGIC_VECTOR(3 DOWNTO 0);
  D  : IN  STD_LOGIC_VECTOR(7 DOWNTO 0);
  P  : IN  STD_LOGIC;
  TX : OUT STD_LOGIC
  );
END MUX_SALIDA;

ARCHITECTURE SELECCION OF MUX_SALIDA IS
BEGIN
  PROCESS(M,D,P)
  BEGIN
    CASE M IS
      WHEN "0000" => TX <= '0';
      WHEN "0001" => TX <= D(0);
      WHEN "0010" => TX <= D(1);
      WHEN "0011" => TX <= D(2);
      WHEN "0100" => TX <= D(3);
      WHEN "0101" => TX <= D(4);
      WHEN "0110" => TX <= D(5);
      WHEN "0111" => TX <= D(6);
      WHEN "1000" => TX <= D(7);
      WHEN "1001" => TX <= P;
      WHEN OTHERS => TX <= '1';
    END CASE;
  END PROCESS;
END SELECCION;
```

Code 5. MAQUINA_FSM.vhd.

```
LIBRARY IEEE;
USE IEEE.STD_LOGIC_1164.ALL;

ENTITY MAQUINA_FSM IS
  PORT(
  RST : IN  STD_LOGIC;
  CLK : IN  STD_LOGIC;
  B   : IN  STD_LOGIC;
  STT : IN  STD_LOGIC;
  EOT : OUT STD_LOGIC;
  M   : OUT STD_LOGIC_VECTOR(3 DOWNTO 0)
  );
END MAQUINA_FSM;

ARCHITECTURE CONTROL OF MAQUINA_FSM IS
SIGNAL QN, QP : STD_LOGIC_VECTOR(3 DOWNTO 0);
BEGIN
  COMBINACIONAL:PROCESS(QP,STT,B)
  BEGIN
    CASE QP IS
        WHEN "0000" =>  IF (STT = '0') THEN QN <= QP;
                        ELSE QN <= "0001";
                        END IF;
                        EOT <= '1'; M <= "1111";
        WHEN "0001" =>  IF (B = '0') THEN QN <= QP;
                        ELSE QN <= "0010";
                        END IF;
                        EOT <= '0'; M <= "1111";
        WHEN "0010" =>  IF (B = '0') THEN QN <= QP;
                        ELSE QN <= "0011";
                        END IF;
                        EOT <= '0'; M <= "0000";
        WHEN "0011" =>  IF (B = '0') THEN QN <= QP;
                        ELSE QN <= "0100";
                        END IF;
                        EOT <= '0'; M <= "0001";
```

```
                    WHEN "0100" =>  IF (B = '0') THEN QN <= QP;
                                    ELSE QN <= "0101";
                                    END IF;
                                    EOT <= '0'; M <= "0010";
                    WHEN "0101" =>  IF (B = '0') THEN QN <= QP;
                                    ELSE QN <= "0110";
                                    END IF;
                                    EOT <= '0'; M <= "0011";
                    WHEN "0110" =>  IF (B = '0') THEN QN <= QP;
                                    ELSE QN <= "0111";
                                    END IF;
                                    EOT <= '0'; M <= "0100";
                    WHEN "0111" =>  IF (B = '0') THEN QN <= QP;
                                    ELSE QN <= "1000";
                                    END IF;
                                    EOT <= '0'; M <= "0101";
                    WHEN "1000" =>  IF (B = '0') THEN QN <= QP;
                                    ELSE QN <= "1001";
                                    END IF;
                                    EOT <= '0'; M <= "0110";
                    WHEN "1001" =>  IF (B = '0') THEN QN <= QP;
                                    ELSE QN <= "1010";
                                    END IF;
                                    EOT <= '0'; M <= "0111";
                    WHEN "1010" =>  IF (B = '0') THEN QN <= QP;
                                    ELSE QN <= "1011";
                                    END IF;
                                    EOT <= '0'; M <= "1000";
                    WHEN "1011" =>  IF (B = '0') THEN QN <= QP;
                                    ELSE QN <= "1100";
                                    END IF;
                                    EOT <= '0'; M <= "1001";
                  WHEN OTHERS =>    QN  <= "0000";
                                    EOT <= '0'; M <= "1111";
                END CASE;
              END PROCESS COMBINACIONAL;

              SECUENCIAL:PROCESS(RST,CLK)
              BEGIN
                IF(RST = '1')THEN
                  QP <= (OTHERS => '0');
                ELSIF(CLK'EVENT AND CLK = '1')THEN
                  QP <= QN;
                END IF;
              END PROCESS SECUENCIAL;
            END CONTROL;
```

Code 6. PARIDAD.vhd.

```
LIBRARY IEEE;
USE IEEE.STD_LOGIC_1164.ALL;

ENTITY PARIDAD IS
  PORT(
  D : IN  STD_LOGIC_VECTOR(7 DOWNTO 0);
  P : OUT STD_LOGIC
  );
END PARIDAD;

ARCHITECTURE SIMPLE OF PARIDAD IS
BEGIN
  P <= ((D(0) XOR D(1)) XOR (D(2) XOR D(3))) XOR ((D(4) XOR D(5)) XOR (D(6) XOR D(7)));
END SIMPLE;
```

4.4 Bluetooth and XBee modules

The wireless modules are configured through AT commands. Command strings have the form ATxx (where xx is the name of a setting). The mode for both is slave to receive data from UART driver architecture. Bluetooth can be set to baud rate from 9600 to 921,600 bps. XBee can be set to baud rate from 9600 to 250,000 bps. Terminal software like Tera Term [26] can be used to have an initial configuration of the devices. Any USB to TTL converter, for example, PL2303HX device or similar, can be used.

In the case of Bluetooth, the module only needs to be connected to the Rx of module to Tx of USB-TTL converter and Tx of module. It is necessary to connect ground and Vcc. HC-06 module is permanently configured to be slaved, and it is always in AT mode when not paired to any other device. AT commands can be founded in datasheets [27].

XBee configuration needs a test utility (XCTU) to enable users to interact with radio frequency (RF) devices through a graphical interface. The application includes built-in tools that make it easy to set up, configure, and test RF devices. The software can be downloaded from the Internet [28].

4.5 Wi-fi driver

ESP8266 Wi-Fi module is used to transmit the sensor data wirelessly to the Wi-Fi modem at the other end with Internet connection. ESP8266 can be initialized using a set of AT commands. Initialization process includes (a) verifying the communication between ESP8266 module and FPGA architecture (RST command) and (b) searching for a Wi-Fi network within its range and connecting to it, with the required credentials (CWJAP command). Sending process includes (a) setting the Wi-Fi module as a TCP/IP client (CIPSTART command); (b) transmitting data involves communication with cloud server using IP address (CIPSEND command). Address IP of the server is required to access the data from personal computing devices such laptop, tablet, and smartphone. **Figure 16** shows the AT command sequence and a block of Wi-Fi architecture:

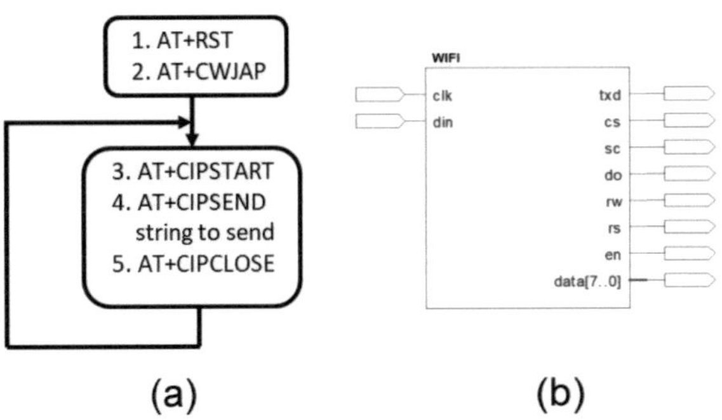


```
LIBRARY IEEE;
USE IEEE.STD_LOGIC_1164.ALL;
USE IEEE.STD_LOGIC_ARITH.ALL;
USE IEEE.STD_LOGIC_UNSIGNED.ALL;

ENTITY WIFI IS
    PORT (
      CLK    : IN  STD_LOGIC;
      TXD    : OUT STD_LOGIC;
      CS     : OUT STD_LOGIC;
      SC     : OUT STD_LOGIC;
      DO     : OUT STD_LOGIC;
      DIN    : IN  STD_LOGIC;
      RW     : OUT STD_LOGIC;
      RS     : OUT STD_LOGIC;
      EN     : OUT STD_LOGIC;
      DATA   : OUT STD_LOGIC_VECTOR(7 DOWNTO 0));
END WIFI;

ARCHITECTURE BEHAVIOR OF WIFI IS

TYPE ARR IS ARRAY (1 TO 167) OF STD_LOGIC_VECTOR(7 DOWNTO 0);

CONSTANT STR: ARR:= ( X"41",X"41",X"54",X"2B",X"52",X"53",X"54",X"0D",X"0A",        -- ATT+RST <CR+LF>
X"41",X"54",X"2B",X"43",X"57",X"4A",X"41",X"50",X"3D",                              -- AT+CWJAP=
X"22",X"4C",X"61",X"62",X"5F",X"49",X"45",X"45",X"5F",X"69",X"6E",X"76",X"69",X"74",X"61",X"64",X"6F",X"22",X"2C",   -- "Lab_IEE_invitado",
X"22",X"45",X"4C",X"45",X"43",X"54",X"52",X"49",X"43",X"41",X"22",X"0D",X"0A",      -- "ELECTRICA" <CR+LF>
X"41",X"54",X"2B",X"43",X"49",X"50",X"53",X"54",X"41",X"52",X"54",X"3D",X"22",X"54",X"43",X"50",X"22",X"2C",   -- AT+CIPSTART="TCP",
X"22",X"31",X"38",X"34",X"2E",X"31",X"30",X"36",X"2E",X"31",X"35",X"33",X"2E",X"31",X"34",X"39",X"22",X"2C",X"38",X"30",X"0D",X"0A", -- "184.106.153.149",80
X"41",X"54",X"2B",X"43",X"49",X"50",X"53",X"45",X"4E",X"44",X"3D",X"34",X"39",X"0D",X"0A",   -- AT+CIPSEND=49 <CR+LF>
X"47",X"45",X"54",X"20",X"2F",X"75",X"70",X"64",X"61",X"74",X"65",X"3F",X"61",X"70",X"69",X"5F",X"6B",X"65",X"79",X"3D",-- GET /update?api_key=
X"51",X"48",X"53",X"36",X"45",X"34",X"48",X"35",X"57",X"34",X"4D",X"42",X"55",X"39",X"5A",X"34",X"26",X"66",X"69",X"65",X"6C",X"64",X"31",X"3D",X"31",X"0D",X"0A'
X"41",X"54",X"2B",X"43",X"49",X"50",X"43",X"4C",X"4F",X"53",X"45",X"0D",X"0A");   -- AT+CIPCLOSE <CR+LF>
```

(c)

Figure 16.
AT command sequence. (a) Flow diagram of WEB connection [27]. (b) Block WI-FI driver module. (c) Code for AT command definition inside FPGA; green letters are comments about hexadecimal data.

5. Experimental results on graphical user interface (GUI)

An analog signal is generated by the function generator to test the system, and the final data sent to the PC or WEB page is observed. **Figures 17**, **18** and **19** show the corresponding practical wave and storage wave.

The GUI (**Figure 18**) was made using Java Eclipse Oxygen [29] and serial communication libraries (jSerialComm). jSerialComm is a Java library designed to provide a platform-independent way to access standard serial ports without requiring external

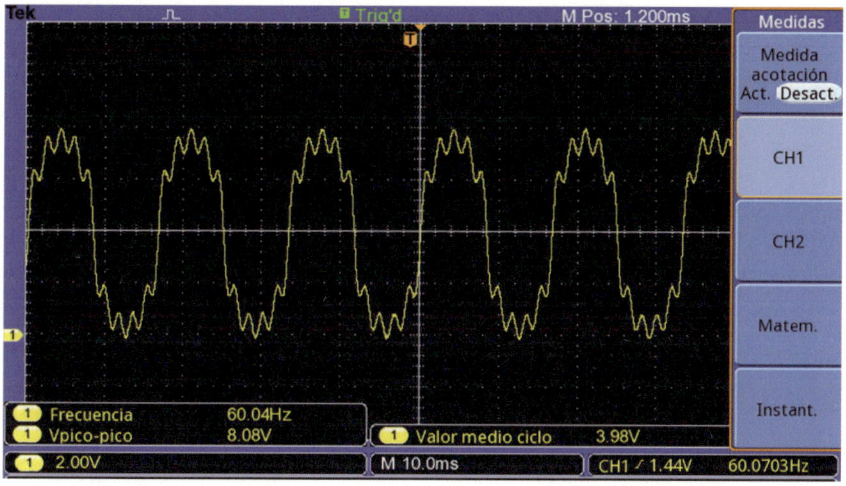

Figure 17.
Measurements of real signal sent to the host and WEB page. The signal has an offset = 3.98Vdc and 8.06Vpp and frequency of 60 Hz with harmonics of 3rd, 5th, 7th, and 9th. This signal is obtained from digital oscilloscope.

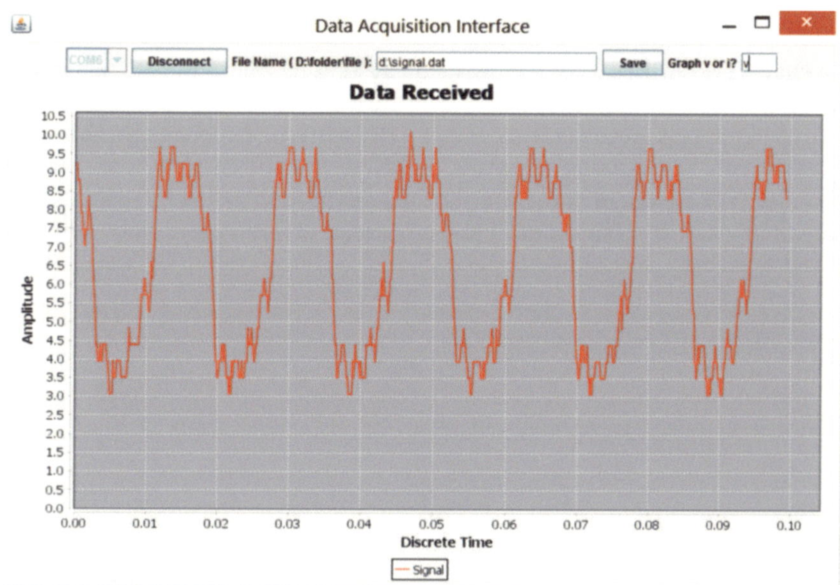

Figure 18.
Data received from the remote DAQ system (Bluetooth or XBee module) using GUI development. Each cycle is represented by 133 samples (sampling frequency = 8 kHz). The UART baud rate is 115,200 bps.

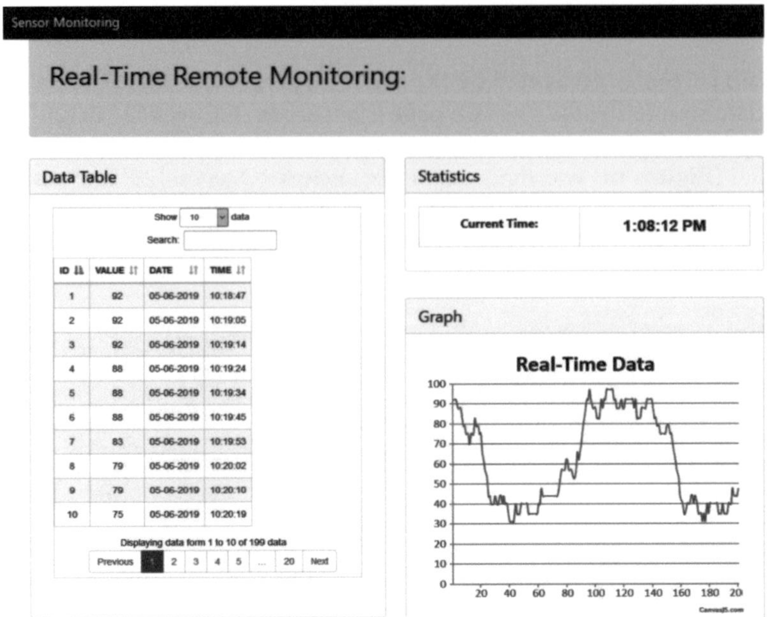

Figure 19.
Data sent to WEB page through Wi-fi module. The WEB page was made on a XAMPP package that includes apache WEB server, MySQL, and PHP [31].

libraries, native code, or any other tools. It is meant as an alternative to Rx-Tx and the (deprecated) Java Communications API, with increased ease of use, an enhanced support for timeouts, and the ability to open multiple ports simultaneously [30].

6. Conclusions

This chapter described a data acquisition system based on FPGA. Several architectures to ADC controller, UART communication, FIFO memory, and Wi-Fi configuration process were made to develop the system. Experiments show that the system can convert the analog signals to digital signal and send to host computer to Java GUI or WEB page in real-time. The data can be acquired by using custom sampling frequency and baud rate. The entire system is designed to be simple, stable, and low cost.

Author details

J. Guadalupe Velásquez-Aguilar*, Outmane Oubram and Luis Cisneros-Villalobos
Faculty of Chemical Sciences and Engineering, Department of Electrical Engineering, Autonomous University of the State of Morelos, Cuernavaca, Morelos, México

*Address all correspondence to: jgpeva@uaem.mx

References

[1] Siewert S, Pratt J, editors. Real-time embedded components and system with linux and RTOS. Mercury Learning and Information LLC; 2016. 483 p. ISBN: 978-1942270041

[2] Mohit A editor. Embedded system design: Introduction to SoC system architecture. Learning Bytes Publishing; 2016. 214 p. ISBN: 978-0997297201

[3] Rajsuman R editor. System-on-a-Chip: Design and Test. Artech House; 2000. 294 p. ISBN: 978-1580531078

[4] Jerraya AA, Mint WW. Hardware/software interface codesign for embedded systems. Computer, IEEE. 2005;**38**:63-69

[5] Kirianaki N, Yurish S, Shpak N, Deynega V. Data Acquisition and Signal Processing for Smart Sensors. John Wiley & Sons Ltd; 2002. 291 p. ISBN: 0-470843179

[6] Velásquez-AguilarJG,Aquino-RobleroF, Limón-MendozaM,Cisneros-VillalobosL, Zamudio-Lara A. Multi-channel data acquisition and wireless communication FPGA-based system, to real-time remote monitoring. In: 2017 International Conference on Mechatronics, Electronics and Automotive Engineering (ICMEAE); 21-24 November 2017; Cuernavaca, México. IEEE; 2017. pp. 181-186

[7] Collotta M, Pau G, Salerno VM, Scatá G, editors. Wireless Sensor Networks to Improve Road Monitoring. IntechOpen; 2012. pp. 323-346. DOI: 10.5772/48505.ch15

[8] National Instruments. Introduction to Bluetooth Device Testing: From Theory to Transmitter and Receiver Measurements. Available from: http://download.ni.com/evaluation/rf/intro_to_bluetooth_test.pdf

[9] Symmetry Electronics, Bluetooth 1.0 vs 2.0,vs 3.0 vs 4.0 vs 5.0—How They Compare. Available from: https://www.semiconductorstore.com/blog/2018/Bluetooth-1-0-vs-2-0-vs-3-0-vs-4-0-vs-5-0-How-They-Differ-Symmetry-Blog/3147

[10] Gupta NK, editor. Inside Bluetooth Low Energy. Artech House; 2016. 458 p. ISBN: 978-1630810894

[11] National Instruments. The Basic of ZigBee Transmitter Testing. Available from: www.ni.com

[12] Jaiswal L, Kaur J, Singh G. Performance analysis of backoff exponent behaviour at MAC layer in ZigBee sensor networks. International Journal of Computer Applications. 2012;**57**(22)

[13] Cunha A, Koubâa A, Severino R, Alves M. Open-ZB: An open-source implementation of the IEEE 802.15.4/ZigBee protocol stack on TinyOS. Available from: https://www.cister.isep.ipp.pt/docs/open_zb__an_open_source_implementation_of_the_ieee_802_15_4_zigbee_protocol_stack_on_tinyos/381/view.pdf

[14] Tennina S et al. editors. IEEE 802.15.4 and ZigBee as enabling Technologies for Low-Power Wireless Systems with Quality-of-Service Constraints. Springer; 2013. 173 p. DOI: 10.1007/978-3-642-37368-8

[15] Available from: https://www.wi-fi.org

[16] Wi-Fi Alliance. Generational Wi-Fi User Guide. Available from: https://www.wi-fi.org/file/generational-wi-fi-user-guide

[17] Rabbit Web site. An Introduction to Wi-Fi. Available

from: http://ftp1.digi.com/support/documentation/0190170_b.pdf

[18] Terasic, DE0-Nano-SoC User Manual. Available from: https://media.digikey.com/pdf/Data%20Sheets/Terasic%20Technologies/DE0-Nano-SoC_UM.pdf

[19] LTC2308 Datasheets, Linear Technology Corporation, 2007

[20] Cyclone V. Hard Processor System Technical Reference Manual, Intel FPGA; 2018. cv_5v4 | 2019.06.14

[21] HC-06 Datasheets, Guangzhou HC Information Technology Co. Ltd. 2011. Available from: https://www.olimex.com/Products/Components/RF/BLUETOOTH-SERIAL-HC-06/resources/hc06.pdf

[22] Espressif. ESP8285 Datasheet. Available from: https://www.espressif.com/sites/default/files/documentation/0a-esp8285_datasheet_en.pdf

[23] Altera, Using the DE0-Nano ADC Controller. Available from: ftp://ftp.intel.com/Pub/fpgaup/.../Using_DE0-Nano_ADC.pdf

[24] Stroud CE. First-In First-Out (FIFO) Control Logic VHDL Modeling Example, ECE Department, Auburn University. Available from: http://www.eng.auburn.edu/~strouce/class/elec4200/vhdlmods.pdf

[25] Digi International Inc. XBee-PRO 900/ DigiMesh 900 OEM RF Modules. Available from: https://www.sparkfun.com/datasheets/Wireless/Zigbee/XBee-900-Manual.pdf

[26] Available from: https://tera-term.en.lo4d.com/windows

[27] Espressif. ESP8266 AT Command Examples. 2017. Available from: https://espressif.com/sites/default/files/.../4b-esp8266_at_command_examples_en.pdf

[28] Available from: https://www.digi.com/products/iot-platform/xctu#productsupport-utilities

[29] Available from: https://www.eclipse.org/

[30] Available from: https://fazecast.github.io/jSerialComm/

[31] Available from: https://sourceforge.net/projects/xampp/

Chapter 2

Real-Time Echo State Network Based on FPGA and Its Applications

Yongbo Liao

Abstract

In this chapter, a hardware processing architecture of real-time echo state network based on field-programmable gate array (FPGA) is proposed, which solves the problem that it is difficult to obtain the output weight of the network in real time. The design of this architecture strictly follows the reservoir calculation (RC) theory, and its five components are established in FPGA: input module, reservoir module, output module, training module, and system switch module. This paper implements the architecture in Altera FPGA chip and verifies it through the application of pattern recognition, waveform generation, and multiple-input multiple-output (MIMO) channel prediction. Experimental results show that the hardware-implemented real-time echo state network can identify the duty cycle of different input signals, generate floating-point waveforms, and predict the MIMO channel by training. In this paper, a real-time echo state network based on field programmable gate array is proposed, which has the advantages of fast computation speed, less resource consumption, and ideal simple task execution.

Keywords: FPGA, ESN, pattern recognition, waveform generation, MIMO channel prediction, real-time, training, testing

1. Introduction

Echo state network [1] simplifies training tasks into linear regression tasks. It mainly solves the problems of large consumption of Recurrent Neural Network (RNN) training resources, long running time, and slow convergence. There are many studies on the applications of echo state network, such as wind power ramp time prediction [2, 3], medical image recognition classification [4], water flow prediction [5], etc. There are also many studies on the structure of the echo state network, such as the dynamic reservoirs that increase their stochastic properties [6] or delay characteristics [7], correlation entropy replaces traditional error function [8], change calculation model [9, 10], etc. Less research work on hardware platform implementation of neural networks, such as [11] proposed a software framework for simulating RNN circuits, [12, 13] proposed FPGA/software framework; however these frameworks are always trained in software such as MATLAB, which not be strictly said to be hardware implementation. The FPGA-based real-time echo state network structure proposed in this chapter trains the output weights on the FPGA platform without calculating the relevant parameters by means of software. In order to verify the performance of the proposed architecture, two types of benchmark tasks were performed: the output signal which was a binary signal [14] and a floating point number signal and a MIMO channel prediction task [15].

2. Theory and model

This section describes the mathematical model of the echo state network. The structural model is shown in **Figure 1**. Let the model have K input units whose vector form is

$$u(n) = (u_1(n), u_2(n)\cdots\cdots\cdots u_K(n))^T \tag{1}$$

N reservoir units, the vector form is

$$x(n) = (x_1, x_2\cdots\cdots\cdots x_N)^T \tag{2}$$

L output units whose vector form is

$$y(n) = (y_1, y_2\cdots\cdots\cdots y_L)^T \tag{3}$$

where $(\bullet)T$ is the transpose, n is the discrete time, and the input/reservoir/output connection weight is represented by a weight matrix of size N × K/N × N/L×(K + N), i.e.

$$W^{in} = \left(w_{i,j}^{i,n}\right), W = (w_{i,j}), W^{out} = \left(w_{ij}^{out}\right) \tag{4}$$

The output unit can select whether to feed back to the intermediate unit and the connection weight is represented by a feedback weight matrix of size N × L:

$$W^{\text{back}} = \left(w_{ij}^{back}\right) \tag{5}$$

Intermediate cell status updates according to the formula

$$x(n+1) = f\left(W^{in}u(n+1) + Wx(n) + W^{back}y_{target}(n)\right) \tag{6}$$

where u(n + 1) is the external given input at time n + 1, such as Eq. (1); $y_{\text{target}}(n)$ is the ideal output at time n, in the form of Eq. (2); and f represents the transfer

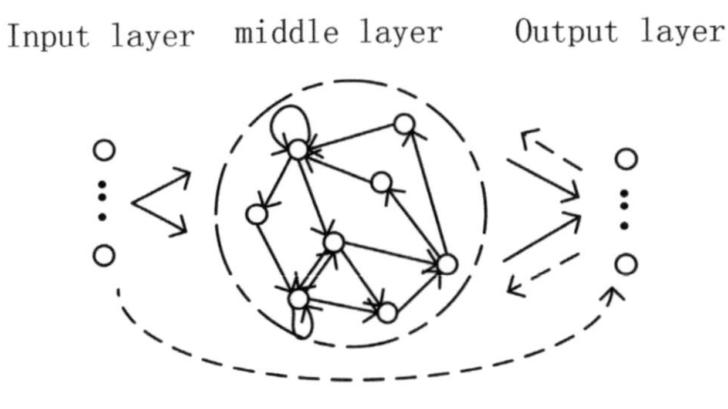

Figure 1.
The basic structure of the echo state network. The arrows indicate the flow of data.

function of the intermediate unit, mainly using the S function, but sometimes a linear network $f = 1$ is also used. Output calculation is based on

$$y(n+1) = f_{out}(W^{out}(u(n+1), x(n+1))) \tag{7}$$

where $(u(n + 1), x(n + 1))$ represents the juxtaposition of the input and the intermediate state vector, as shown in the input layer to the output layer of the dashed arrow in **Figure 1**, in some applications, such as [16], where the data stream does not exist. That is, the output is calculated directly using the intermediate state value. The output transfer function is usually f_{out} = tanh or $f_{out} = 1$, depending on whether the output unit is nonlinear or linear. The output weight W^{out} is calculated according to the following formula:

$$W^{out} = Y_{target}X^T\left(XX^T + \alpha^2 I\right)^{-1} \tag{8}$$

where $I \in R^{N\times N}$ is the identity matrix, α is the regularization factor, $R \in R^{N\times N}$ is the set matrix of $(u(n+1), (x(n+1)))$, Y_{target} is the ideal output set matrix, and $(\bullet)^{-1}$ is the matrix inversion.

3. Real-time FPGA echo state network structure

Real-time FPGA echo state network execution structure maps Eqs. (6)–(8) to six modules, which are input module, reservoir module, output module, training module, system switch module, and random number generator, as shown in **Figure 2**. The input module is a two-input single-output module, and the input is a random input weight Win generated by a random number generator and an external signal $u(n + 1)$, performing a W in $u(n + 1)$ multiplication operation, and encoding the input signal to form a data signal that can be calculated by the reservoir module. The reservoir module is a five-input single-output module that strictly performs the remainder of Eq. (6), the input including the encoded external signal from the input module, reservoir initial state value (this input is the state value of the previous clock reservoir module output as the operation progresses), network expected output, reservoir interconnection weights generated by the random number generator, and feedback. Connecting the weight, output high-dimensional state signal, the specific circuit is shown in **Figure 3**.

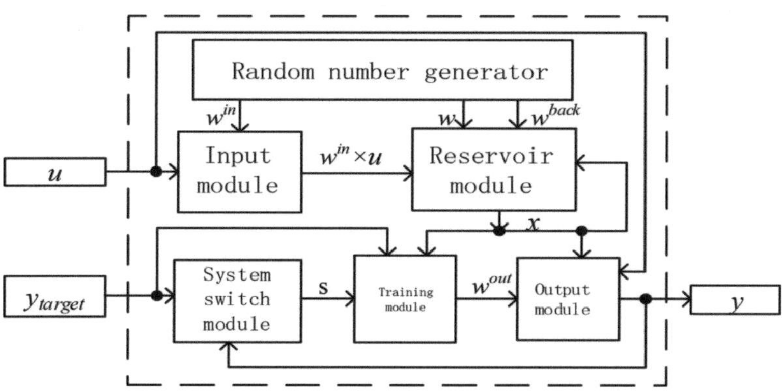

Figure 2.
Real-time FPGA echo state network structure diagram.

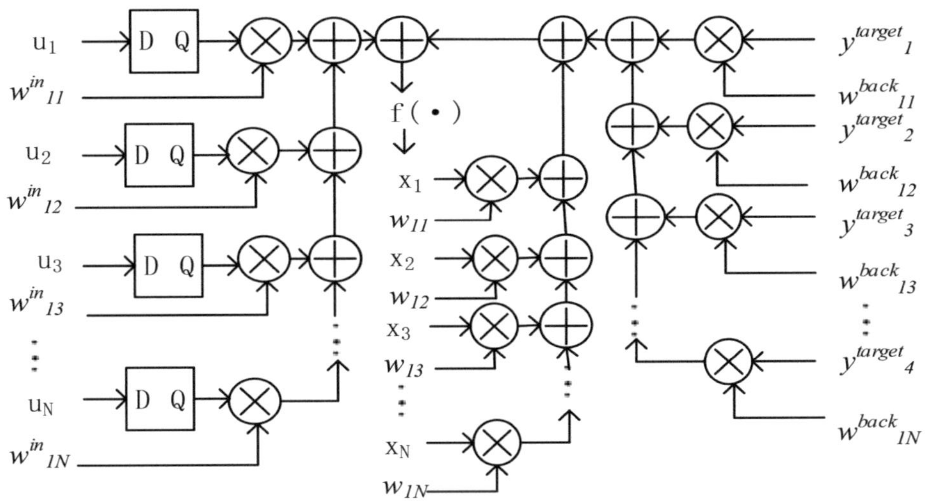

Figure 3.
Reservoir module.

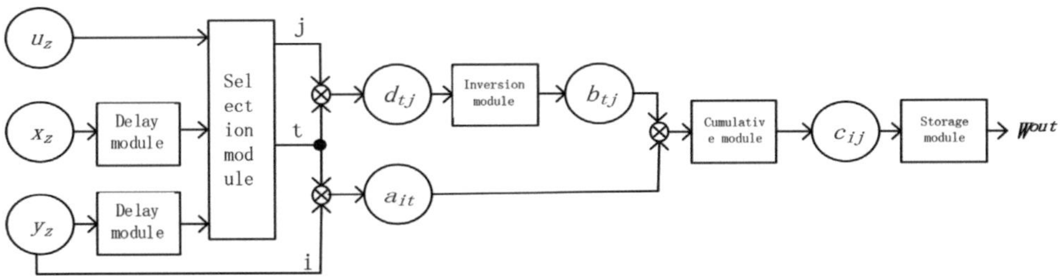

Figure 4.
Training module. Subscript z represents the zth unit of input, reservoir, and output.

(**Figure 3** circuit is the circuit that outputs the reservoir state x1, other state circuits are similar). The output module is a three-input single-output module, the input is an external input signal, the state value is obtained by the reservoir module calculation, and the output connection weight is calculated by the training module. The output is the final acquired network output. The function is to perform simple multiplication and addition on the input data and decode the result to form an output signal. The training module is a three-input single-output module, the input is an externally given desired output, a status signal is generated by the reservoir module, and an enable signal S is sent by the system switch module, When $S = 0$, the module stops working. When $S = 1$, the module works, and the output is the core parameter output connection weight of the echo state network. The function is expressed as Eq. (8). The specific implementation mechanism is shown in **Figure 4** (in Eq. (8)), $Y_{target}X^T = (a_{it})$, $XX^T + \alpha^2 I = (d_{tj})$, $(XX^T + \alpha^2 I)^{-1} = (b_{tj})$, $W^{out} = (c_{ij})$; then, the output weight $W^{out} = (c_{ij})$ is obtained; The system switch module is a two-input single-output module. The input is a network output signal and a desired output signal. The output is an enable signal for controlling the operation of the training module. Its function is mainly to judge the network performance. When the network output signal matches the expected output signal or the difference is within the receiving range, the output $S = 0$ and the other cases continue to output $S = 1$. A random number generator is used to generate inputs, reservoirs, and feedback weights.

The following sections detail how to train a real-time FPGA echo state network. As is well known, FPGAs implement digital systems that primarily process digital signals, providing a logic cell array that can be configured as a given function via a bitstream file. Its basic digital logic, the smallest programmable logic unit, is the logic gate. Therefore, FPGAs are the best device for performing echo state networks. The architecture is all implemented in the FPGA. On the one hand, the dynamic reservoir (i.e., the middle layer) is established. On the other hand, how to obtain the real-time output weights when the input signal is the digital signal "0" or "1" is established.

The dataflow and training process of the real-time FPGA echo state network structure will be briefly described as follows:

Given: Input and target output sequence $u(n)$ and $y_{\text{target}}(n)$, $\text{n} = 1\cdots\cdots T$.

Objective: The teacher signal input/output training acquires W^{out} and acquires the network output signal by loading the input signal.

Proceed as follows:

The random number generator module generates an input, a reservoir, and a feedback weight of the echo state network and sends the input and feedback weights to the input module, and the reservoir weight is sent to the reservoir module.

The input module loads the input signal, encodes the input signal, and sends it to the reservoir module.

The reservoir module receives the encoded signal sent by the input module, loads the target output signal, acquires the feature value, and sends it to the output module and the training module.

The training module loads the input signal, the target output signal, and the characteristic value sent by the reservoir module; calculates the output weight; sends the result to the output module; and determines whether to stop training according to the signal sent by the system switch module.

The output module loads the input signal and calculates the network output according to the characteristic value sent by the reservoir module and the output weight sent by the training module and sends the network output value to the system switch module and outputs the actual network value.

The system switch module loads the target output signal and the network output signal to determine whether the match is matched and sends the judgment result back to the training module. If it matches, it sends back $S = 0$; if it does not match, it sends back $S = 1$;

Repeat steps 2–6 until the judgment result of the system switch module is to stop training, that is, $S = 0$.

This is followed by the regular echo state network function, which only performs input, reservoir, and output modules and outputs network prediction values.

4. Experiment and analysis

A total of two types of benchmark task experiments were performed: the output signal was a binary signal and a floating point number signal and a multiple-input multiple-output channel prediction task experiment. The programming language is Verilog HDL, the chip uses Altera Stratix III FPGA, and the integrated place and route are implemented in QUARTUS II.

4.1 Different duty cycle signal pattern recognition

The pattern recognition reference task with different duty cycles is very similar to the memory resistance based on reservoir calculation (RC) in [14]. The mode signals with different duty cycles are shown in **Figure 5**. The input signal of the echo

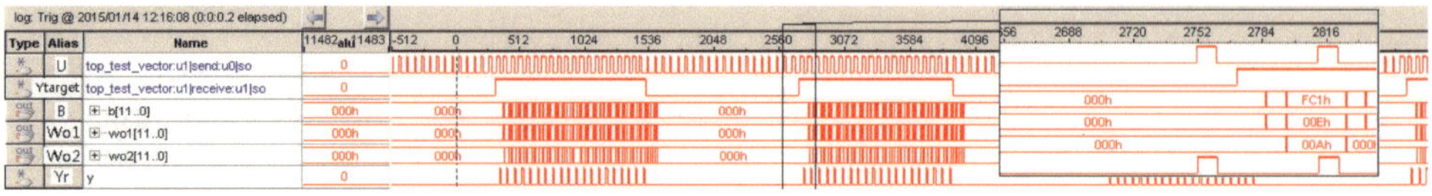

Figure 5.
Different duty cycle pattern recognition training.

state network is *U*, the expected output is *Y* target, and the actual output signal is Yr; Wo1 and Wo2 are output weights, and B is the bias signal. The second line (signal Y_target) represents the expected response of the first line (signal U). When the duty cycle of the input signal is less than 50%, the signal Y_target should converge to 0 and should converge to 1 for a duty cycle greater than 50%. As shown in **Figure 5**, the online training echo state network in the FPGA obtains Wo1, Wo2, and B, and their values are 00Eh, 00Ah, and FC1h, respectively. After the output weight is obtained, different duty cycle mode signals are loaded into the trained echo state network, and the result is as shown in **Figure 6**. The output signal (Yr) changes between a duty cycle of the input signal (U) greater than 50% and less than 50%.

Figure 7 is a graph showing the percentage of the total number of nerve cells implemented in the FPGA logic and FPGA and the error curve. It can be seen that the logic utilization is less than 60% until the number of neurons is 512 units. When the number of nerves exceeds 16 units, the error between the actual output signal and the ideal output is zero. Therefore, the circuit resources proposed in this paper consume less, and the convergence speed is faster.

4.2 Sine wave generator

Here we test how to train the echo state network to generate a sine wave signal, which is a floating point number, which is very similar to the simple sine wave test

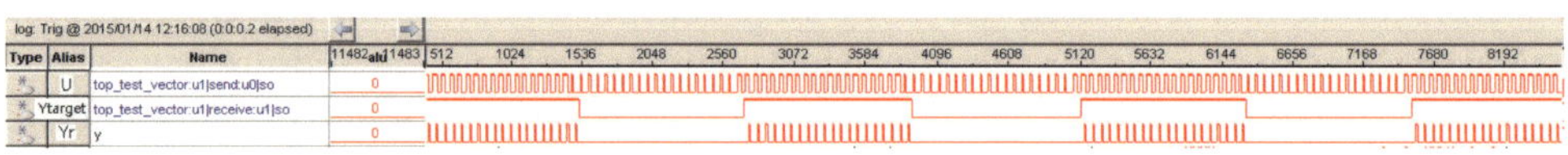

Figure 6.
Echo state network test results.

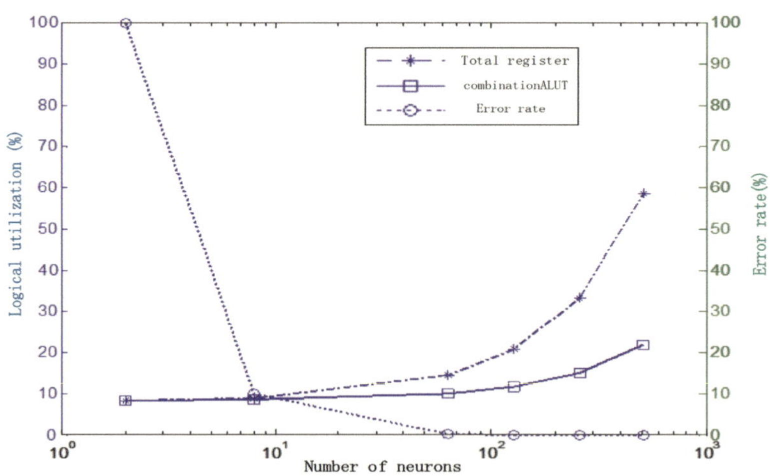

Figure 7.
Relationship between neuron number and logic utilization and error curve.

performed in MATLAB when the echo state network is proposed in [1]. The ideal sine wave signal is given by the equation $y_{target}(t) = \sin(t/8)$. There is no input in this experiment, so set W^{in} to 0; W and W^{back} are matrices formed by random numbers, and the basic unit of the echo state network is the standard S unit (i.e., the activation function is S function tanh), and the training result is shown in **Figure 8**. In the process of generating sine wave training, the ideal output signal (Y_{target}) is a floating point number, so the actual output signal (Yr), the normalized root mean square error signal (E), and the output weight signal (Wout1, Wout2, Wout3, Wout4) are all floating point numbers. The training error is calculated in the system switch module, and E = 3D0228E7h is calculated according to the normalized root mean square error calculation equation $E = \sqrt{\sum_{n=1}^{N} \frac{\left(y_{target}(n) - y(n)\right)^2}{N \times \text{var}\left(y_{target}\right)}}$. The optimized output weights Wout1, Wout2, Wout3, and Wout4 of the training module are BDFEDD9Dh, 3CB22AA8h, 3CA0BDD8h, and 3F55E542h, respectively. The waveform shown in **Figure 9** is acquired by the SignalTap II logic analyzer and is a floating-point sine wave generated for the trained echo state network. The Altera floating-point IP core was used in the experiment to set up the echo state network. As shown in **Figure 10**, the top is the Y target signal, the middle is the actual output y signal of the network, and the bottom is the error waveform of the network expected output and the actual output.

4.3 MIMO channel prediction

Recurrent neural networks have been widely used in MIMO systems [17–23]. Echo state networks are a way to train recurrent neural networks. They have faster convergence characteristics, and more efficient tracking channel state changes than other traditional training methods. For a 2 × 2 multiple-input multiple-output system with a binary phase shift keying (BPSK) modulator, as shown in **Figure 11**, the zero-forcing equalizer used in the receiver section can reduce symbol interference (ISI) due to the precise channel. It is estimated that the zero-forcing equalization can be improved by the degraded radio channel, and therefore the proposed architecture is used for MIMO channel prediction.

The matrix equation of the MIMO system shown in **Figure 11** is given as

$$\begin{bmatrix} y_1 \\ y_2 \end{bmatrix} = \begin{bmatrix} h_{11} & h_{12} \\ h_{21} & h_{22} \end{bmatrix} \begin{bmatrix} x_1 \\ x_2 \end{bmatrix} + \begin{bmatrix} n_1 \\ n_2 \end{bmatrix} \tag{9}$$

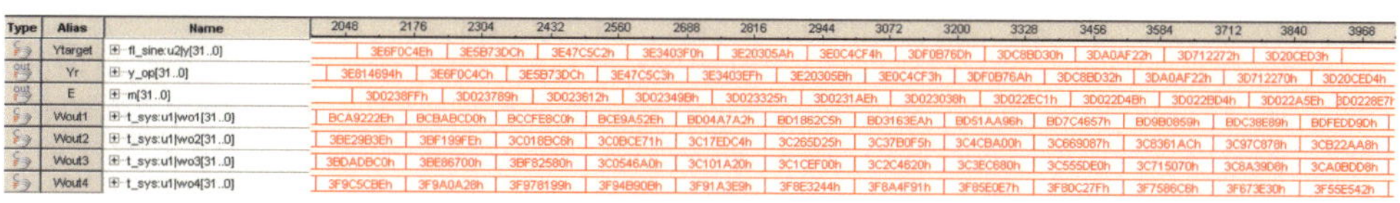

Figure 8.
Sine wave generation training.

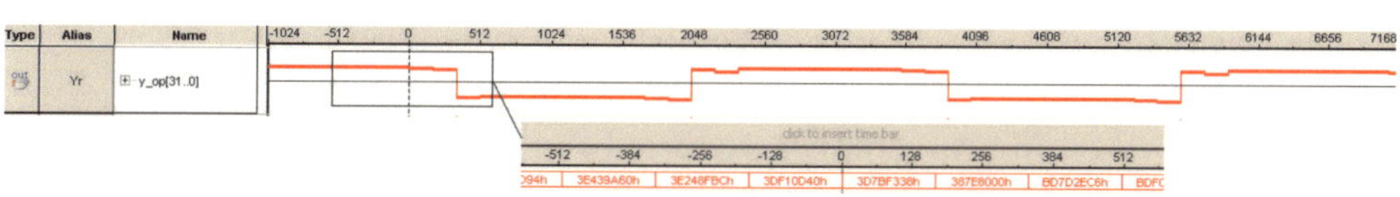

Figure 9.
Echo state network generating floating point sine wave.

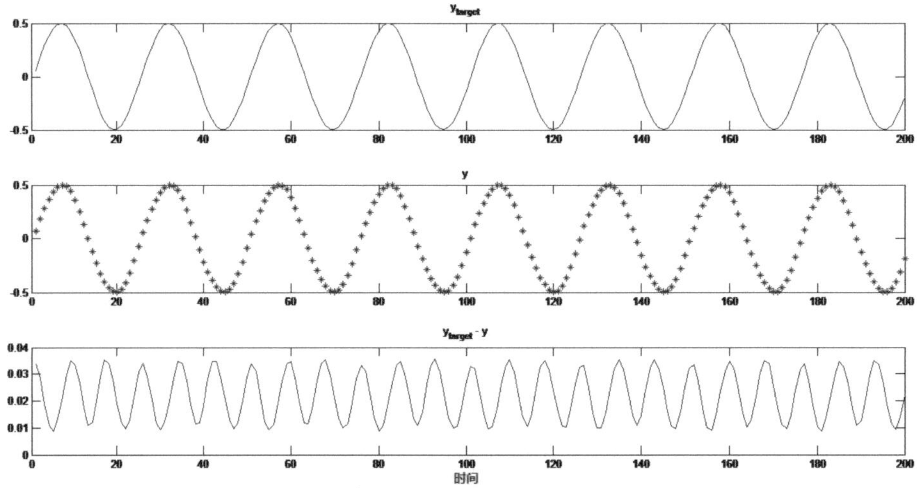

Figure 10.
Target signal (y_{target}), actual signal (y), and error curve (number of neurons is 8).

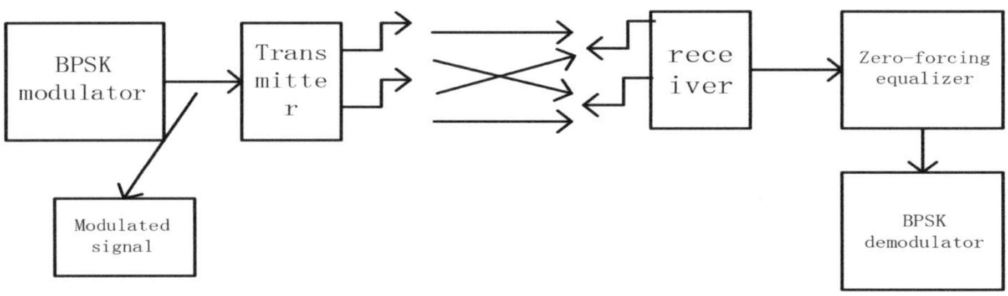

Figure 11.
2 × 2 MIMO system.

The system can be represented as a compact form $Y = \mathrm{HX} + n$, where Y is a 2×1 received signal vector, H is a 2×2 channel coefficient matrix, X is a 2×1 propagation vector, and n is a 2×1 additive white Gaussian noise vector. The channel is considered to be a Rayleigh decay with a mean of 0 and a variance of 0.5. At the receiving end, the zero-forcing equalization performs the prediction of the propagated signal, and the equation is

$$\hat{X} = \left(H^H H\right)^{-1} HY \tag{10}$$

where $\left(H^H H\right)^{-1} H$ represents the pseudo inverse of H. The predicted $\hat{X}$ loaded BPSK demodulator recovers the original information.

In order to dynamically update the channel state at each step, an echo state network is added to the 2×2 MIMO system in **Figure 11**. The system structure diagram after adding the echo state network is shown in **Figure 12**. The echo state network channel prediction strategy is shown in **Figure 13**. The channel coefficients are trained in the echo state network channel prediction. Once the training is completed, the echo state network channel prediction can automatically generate the predicted channel coefficients, and the predicted channel coefficients are loaded into the zero-forcing equalization, thereby completing the MIMO channel prediction.

Figure 14 shows the RTL level circuit diagram generated by the FPGA. The system mainly includes a transmitter (fx), a receiver (rx), and an echo state

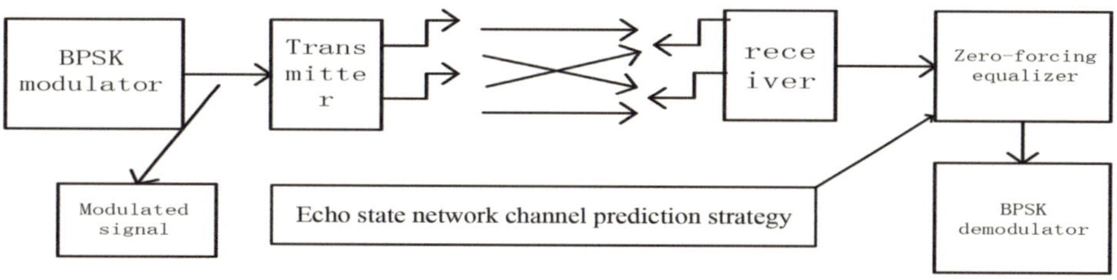

Figure 12.
Echo state network for 2 × 2 MIMO systems.

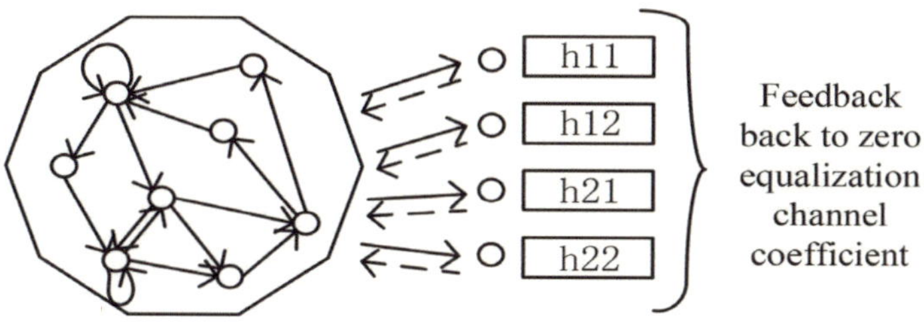

Figure 13.
Echo state network channel prediction strategy structure diagram.

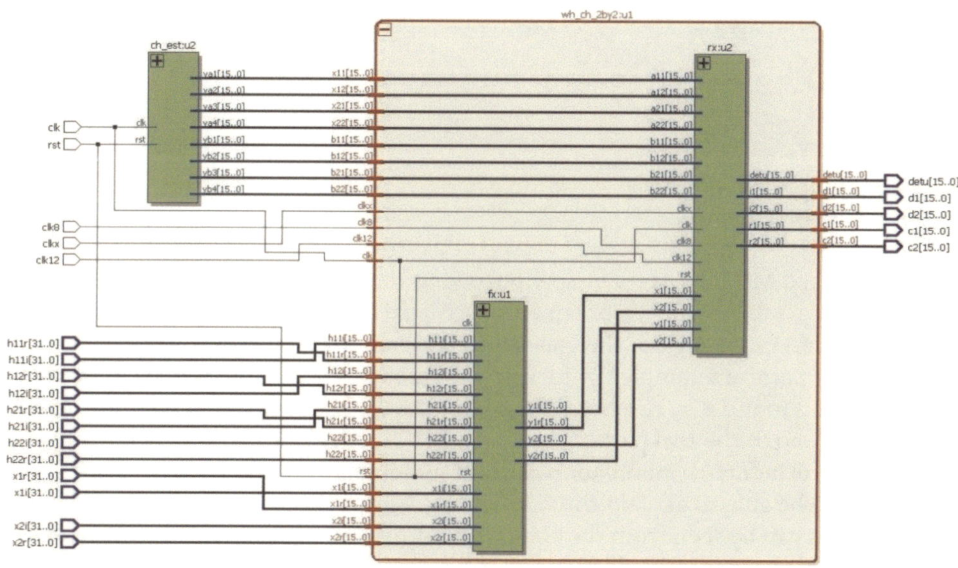

Figure 14.
RTL circuit diagram obtained by FPGA synthesis.

network module (ch_test). In the transmitting module (fx), the signals x1r, x2r and x1i, x2i are the real and imaginary parts of the MIMO system input signals X1 and X2, respectively, and h11r, h12r, h21r, h22r and h11i, h12i, h21i, h22i are, respectively, MIMO channels. The real and imaginary parts of the coefficient, y1r, y2r and y1i, y2i are the real and imaginary parts of the received signal, respectively. In the receiving module (rx), a11, a12, a21, a22 and b11, b12, b21, b22 are the real and imaginary parts of the channel prediction coefficients, and the output is calculated by the echo state network module (ch_test), c1, c2, and d1. d2 is the real and imaginary part of $\hat{X}$, respectively, corresponding to the demodulated signal X, and the signal "detu" is the demodulation scale factor of c1, c2, d1 and d2.

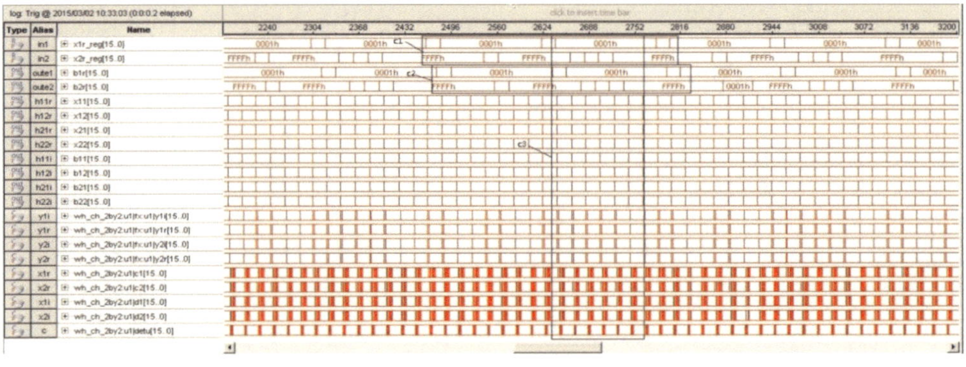

Figure 15.
Real-time waveforms for channel prediction in MIMO systems and echo state networks.

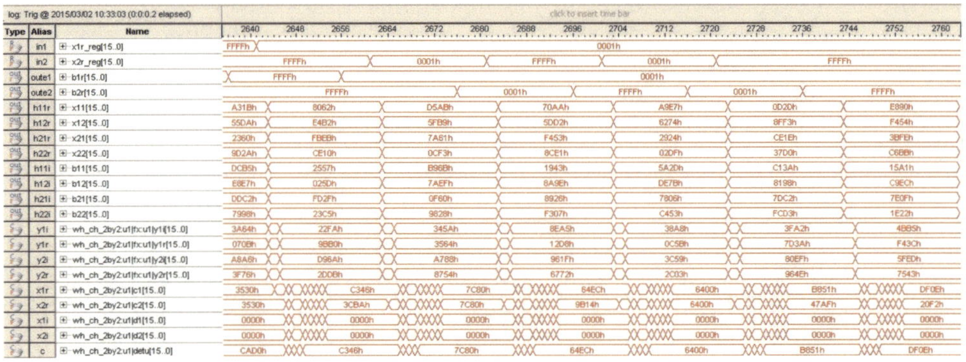

Figure 16.
C3 partial enlargement result.

The designed MIMO system is downloaded to the FPGA chip through a bitstream file, and the waveform result is obtained by a SignalTap II logic analyzer, as shown in **Figure 15**. It can be seen from the waveform diagram that the signal waveforms of the C1 and C2 parts are completely identical. In the C1 portion, the signals in1 and in2 correspond to x1r and x2r, respectively, and in the C2 portion, the signals oute1 and oute2 correspond to the real part of the signal X. X1r, x2r, and x1i, x2i are the real and imaginary parts before demodulation of the demodulated signal, respectively.

In order to be able to explain the C3 part, the C3 part is enlarged here (see **Figure 16**). As can be seen from the C3 section, the received signals Y1 and Y2 and the estimated channel matrix are all changed. However, zero-forcing equalization can still predict values x1r, x2r, and C through the echo state network. After the processing is completed, BPSK demodulation signals "oute1" and "oute2" are obtained.

5. Conclusion

The real-time FPGA echo state network structure is proposed and studied. The input weight and the reservoir weight are randomly determined before training, and the output weight is calculated in real time in the FPGA by training the echo state network. In the above two benchmark experiments (pattern recognition and waveform generation) and MIMO channel prediction experiments, the proposed hardware architecture can recognize the duty cycle of different input signals, generate floating point waveforms, and predict channel coefficients. From the experimental results, the echo state network is faster, the resources are less occupied, and the simple task execution is ideal. In future research work, the proposed FPGA real-time echo state network will be used in more complex 5G-based wireless MIMO-OFDM systems.

Author details

Yongbo Liao
State Key Laboratory of Electronic Thin Films and Integrated Devices, University of Electronic Science and Technology of China, Chengdu, China

*Address all correspondence to: lyb@uestc.edu.com

References

[1] Jaeger H. The "Echo State" Approach to Analysing and Training Recurrent Neural Networks. German National Research Center for Information Technology; 2001

[2] Dorado-Moreno M, Cornejo-Bueno L, Gutiérrez PA, et al. Robust estimation of wind power ramp events with reservoir computing. Renewable Energy. 2017;**111**:428-437

[3] Escalona-Morán MA, Soriano MC, Fischer I, et al. Electrocardiogram classification using reservoir computing with logistic regression. IEEE Journal of Biomedical and Health Informatics. 2015;**19**(3):892-898

[4] Bezerra SGTA, Andrade CBD, Valença MJS. Using reservoir computing and trend information for short-term streamflow forecasting. In: Artificial Neural Networks and Machine Learning —ICANN 2016. Springer International Publishing; 2016

[5] Basterrech S, Rubino G. Echo state queuing networks: A combination of reservoir computing and random neural networks. Probability in the Engineering & Informational Sciences. 2017;**31**:1-20

[6] Pahnehkolaei SMA, Alfi A, Machado JAT. Uniform stability of fractional order leaky integrator echo state neural network with multiple time delays. Information Sciences. 2017; **418-419**:703-716

[7] Guo Y, Wang F, Chen B, et al. Robust echo state networks based on correntropy induced loss function. Neurocomputing. 2017;**267**(6):295-303

[8] Lun SX, Yao XS, Qi HY, et al. A novel model of leaky integrator echo state network for time-series prediction. Neurocomputing. 2015;**159**(1):58-66

[9] Li D, Min H, Wang J. Chaotic time series prediction based on a novel robust echo state network. IEEE Transactions on Neural Networks and Learning Systems. 2012;**23**(5):787-799

[10] Schrauwen B, D'Haene M, Verstraeten D, et al. Compact hardware liquid state machines on FPGA for real-time speech recognition. Neural Networks the Official Journal of the International Neural Network Society. 2008;**21**(2–3):511

[11] Alomar ML, Canals V, Martinez-Moll V, et al. Low-cost hardware implementation of reservoir computers. In: International Workshop on Power and Timing Modeling, Optimization and Simulation. IEEE; 2014. pp. 1-5

[12] Jaeger H. Tutorial on Training Recurrent Neural Networks, Covering BPPT, RTRL, EKF and the Echo State Network Approach - First revision. 2002. 7 p

[13] Sun X, Li T, Li Q, et al. Deep belief echo-state network and its application to time series prediction. Knowledge-Based Systems. 2017;**130**(15):17-29

[14] Yi Y, Liao Y, Wang B, et al. FPGA based spike-time dependent encoder and reservoir design in neuromorphic computing processors. Microprocessors and Microsystems. 2016;**46**(PB):175-183

[15] Jaeger H, Haas H. Harnessing nonlinearity: Predicting chaotic systems and saving energy in wireless communication. Science. 2004; **304**(5667):78-80

[16] Kulkarni MS, Teuscher C. Memristor-based reservoir computing. In: IEEE/ACM International Symposium on Nanoscale Architectures. ACM; 2012. pp. 226-232

[17] Zhang L, Zhang X. MIMO channel estimation and equalization using three-layer neural networks with feedback. Journal of Tsinghua University Natural Science Edition (English Edition). 2007; **12**(6):658-662

[18] Potter C. RNN based MIMO channel prediction. Signal Processing. 2010; **90**(2):440-450

[19] Sarma KK, Mitra A. Modeling MIMO channels using a class of complex recurrent neural network architectures. AEU International Journal of Electronics and Communications. 2012;**66**(4): 322-331

[20] Routray G, Kanungo P. Rayleigh fading MIMO channel prediction using RNN with genetic algorithm. Communications in Computer and Information Science. 2011;**250**:21-29

[21] Cai H. MIMO-OFDM channel estimation based on neural network. Computer Engineering and Applications. 2011;**47**(34):1-4

[22] Lukoševičius M. A Practical Guide to Applying Echo State Networks. Springer; 2012

[23] Dorado-Moreno M, Cornejo-Bueno L, Gutiérrez PA, et al. Robust estimation of wind power ramp events with reservoir computing. Renewable Energy. 2017;**111**:428-437

Chapter 3

Flexible Baseband Modulator Architecture for Multi-Waveform 5G Communications

Mário Lopes Ferreira and João Canas Ferreira

Abstract

The fifth-generation (5G) revolution represents more than a mere performance enhancement of previous generations: it will deeply transform the way humans and/or machines interact, enabling a heterogeneous expansion in the number of use cases and services. Crucial to the realization of this revolution is the design of hardware components characterized by high degrees of flexibility, versatility and resource/power efficiency. This chapter proposes a field-programmable gate array (FPGA)-oriented baseband processing architecture suitable for fast-changing communication environments such as 4G/5G waveform coexistence, noncontiguous carrier aggregation (CA) or centralized cloud radio access network (C-RAN) processing. The proposed architecture supports three 5G waveform candidates and is shown to be upgradable, resource-efficient and cost-effective. Through hardware virtualization, enabled by dynamic partial reconfiguration (DPR), the design space exploration of our architecture exceeds the hardware resources available on the Zynq xc7z020 device. Moreover, dynamic frequency scaling (DFS) enables the runtime adjustment of processing throughput and power reductions by up to 88%. The combined resource overhead for DPR and DFS is very low, and the reconfiguration latency stays two orders of magnitude below the control plane latency requirements proposed for 5G communications.

Keywords: FPGA, reconfigurable computing, dynamic partial reconfiguration, baseband processing, OFDM, FBMC, UFMC, waveform coexistence, carrier aggregation

1. Introduction

The fifth-generation (5G) cellular network technology will have a tremendous impact on society by optimizing existing telecommunication services and applications and enabling solutions in new application fields, such as transportation, education or medical science. The scope of the anticipated changes is clear from the three main types of 5G use cases and services defined by the International Telecommunication Union (ITU): enhanced mobile broadband (eMBB), ultra-reliable and low-latency communications (URLLC) and massive machine-type communications (mMTC) [1]. Therefore, the handling of the physical layer (PHY) for 5G systems will be far more complex than in the current generation.

Orthogonal frequency-division multiplexing (OFDM) is the preferred waveform in 4G standards, and the 3GPP Release 15 [2] recently defined it as the multiple access scheme for the 5G New Radio (NR) PHY, especially due its high frequency selectivity, flexibility, efficient hardware implementation by FFT/IFFT modules, and good Multiple-Input Multiple-Output (MIMO) compatibility [3]. However, the spectrum of OFDM symbols presents large side lobes that cause high out-of-band (OOB) emissions. Moreover, the interference between adjacent time-domain symbols is mitigated by adding redundancy to each symbol, which reduces spectral efficiency. Together, these characteristics may make 5G requirements in certain communication scenarios hard to achieve, which has led to the proposal of other waveforms [4]. The most popular ones are *filter bank multicarrier* modulation (FBMC), *Universal Filtered Multicarrier* modulation (UFMC), Filtered OFDM (f-OFDM) and *generalized frequency-division multiplexing* (GFDM). Different waveforms imply different baseband processing operations. Especially for sub-6 GHz spectrum bands, the *coexistence* of multiple numerologies and waveforms and the close interworking between 5G and current systems is likely to occur in the near future [5].

The expansion of wireless communication caused by 5G systems and services raises concerns about the inefficient use of the electromagnetic spectrum. In addition, to expand spectrum utilization to frequency bands above 6 GHz, a more efficient spectral utilization of heavily used bands must be achieved. To tackle this issue, future baseband processor designs should support *dynamic spectrum access* (DSA) [6] and *carrier aggregation* (CA) schemes.

In summary, baseband processing infrastructures for 5G systems must be (1) *flexible*, to adapt their operation for different communication setups (i.e. waveforms and their parameterization); (2) *scalable*, to tune performance and capacity according to communication demands; (3) resource and power *efficient*, for cost-effectiveness and reduced environmental impact [7]; (4) *forward compatible*, to easily integrate the support for new services and requirements, extending system lifetime. Modern field-programmable gate arrays (FPGAs) represent an implementation platform that favors the design of systems with the characteristics mentioned. The intrinsic FPGA reconfigurability can be enhanced by means of *dynamic partial reconfiguration* (DPR), i.e. by reconfiguring modules of the design without halting the system. The hardware virtualization allowed by DPR enhances system flexibility, feature wealth, upgradability and cost-effectiveness [8]. This chapter discusses how DPR and *dynamic frequency scaling* (DFS) can be combined to produce a dynamically reconfigurable baseband processing architecture for multimode, multi-waveform coexistence and dynamic spectrum aggregation.

After a brief summary of the state of the art in Section 2, the implementation of datapaths for baseband processing of three waveforms (OFDM, FBMC and UFMC) is described in Section 3. The implementation of a dynamically reconfigurable baseband modulator that combines these datapaths is described in Section 4, together with a discussion of the results. Some final remarks are presented in Section 5.

2. Summary of the state of the art

Application of DPR to baseband processing in wireless communications started with the adoption of small-scale and relatively simple functional elements such as FIR filters, constellation mappers or channel encoders [9, 10]. Possibly the first multi-waveform flexible PHY architecture was proposed by He et al. [11]. It is a software-defined radio (SDR) architecture implemented on a Xilinx Virtex-5 FPGA, which combines two reconfiguration techniques: (a) DPR to dynamically change the baseband processing mode of operation (e.g. FFT size, modulation scheme and CP length) and (b) DFS to adapt the clock frequency of the digital up-converter and the baseband processor. The design supports two waveforms (OFDM and WCDMA) and several 3G/4G standards and modes of operation. Compared with a static multimode design, the DPR-based design achieves a reduction in the number of used slices, DSP blocks (DSPs) and block RAMs (BRAMs). However, the comparison is not accurate as the static design uses parallel and independent processing chains for each standard, ignoring potential optimization from the reutilization of common modules.

CoPR, an automated framework for DPR-based adaptive systems on a Xilinx Zynq device, is described in [12]. An illustrative case study is presented, where a reconfigurable multistandard baseband OFDM transmitter is designed. The design supports three standards (IEEE 802.11, IEEE 802.16 and IEEE 802.22) and contains two reconfigurable partitions (RPs): one to implement the digital modulation scheme and the other for the OFDM processing datapath. The paper only reports reconfiguration time results and does not provide figures for power consumption or the amount of resources of each RP.

An ARM-FPGA-based platform is also used in [13]. Several processes run on the ARM processor and retrieve communication environment information, which is employed by a configuration controller to reconfigure an OFDM baseband processing modulator. An OFDM transmitter supporting Wi-Fi and WiMAX is implemented on Zynq's programmable logic (PL) with four RPs used for scrambling, interleaving, FEC encoding and IFFT. Results for resource utilization and DPR latency are discussed, together with power consumption measurements. However, the sampling period used for the measurements is of the order of magnitude of the reconfiguration times (milliseconds) and, therefore, not suitable for accurate real-time measurements at the time scale of interest.

The Zynq is also used in [14], which presents a HW/SW codesign for CR systems combining parameter reconfiguration and DPR. Only DPR latency and RP resources are reported.

Pham et al. [15] present a reconfigurable multistandard OFDM transceiver supporting IEEE 802.11, IEEE 802.16 and IEEE 802.22 on a Xilinx Virtex-6 FPGA. The modulator uses a single RP, whereas the demodulator explores a mixture of DPR-based and static multimode modules. The authors put more weight on the whole architecture and only provide data about reconfiguration times and bitstream size.

All works mentioned target 3G/4G standards and waveforms. From a system perspective, they focus primarily on the enhanced flexibility DPR can offer, with less attention paid to the global impact of this technique on the design of the hardware infrastructure. Additionally, no architecture with multiple and independent processors suitable for noncontiguous spectrum aggregation is studied.

3. Implementation of datapaths for baseband processing

This section describes the implementation of pipelined datapaths for three different waveforms (OFDM, FBMC and UFMC) and respective variants. Each variant is defined by the values assigned to the parameters of the design. The possible sets of values are sometimes called "numerologies" in the literature. In this case study, two sets of parameter values for each waveform are considered, as described in the remainder of this section and summarized in **Tables 1–3**.

Parameter	Mode 1	Mode 2
# subcarriers, N (IFFT size)	512	1024
length of cyclic prefix, L_{CP}	40 (1st slot symbol)	80 (1st slot symbol)
	36 (other symbols)	72 (other symbols)
# WOLA samples, W	4	6

Table 1.
Parameter values supported by the OFDM datapath.

Parameter	Mode 1	Mode 2
# subcarriers, N_c	512	1024
Overlapping factor, K	4	4
IFFT size, $K \times N_c$	2048	4096

Table 2.
Parameter values supported by the FBMC datapath.

Parameter	Mode 1	Mode 2
# subcarriers, N	512	1024
# subcarriers per PRB	12	12
# active PRBs	3	3
IFFT size, N'	64	64
Upsampling factor, N/N'	8	16
Filter length, L	37	73
Filter type	Dolph-Chebyshev (60-dB side lobe attenuation)	

Table 3.
Parameter values supported by the UFMC datapath.

3.1 Baseband datapath for OFDM

OFDM is the main reference among multicarrier modulation waveforms; it is used in a wide range of standards such as DSL, DVB-T, DVB-C, Wi-Fi (IEEE 802.11), WiMAX (IEEE 802.16) and 3GPP LTE. The conceptual structure of an OFDM modulator is illustrated in **Figure 1**.

With exception of the inverse FFT (IFFT), the tasks required for a modulator involve only simple arithmetic, data selection and reordering. The first module is the QAM mapper. For a general M-ary QAM case, the module is simply

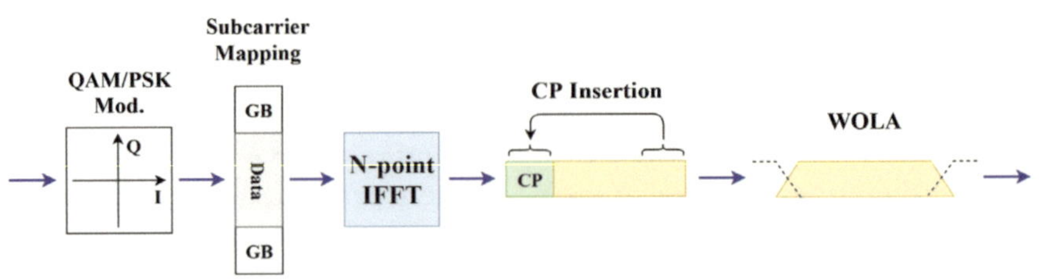

Figure 1.
OFDM baseband modulation. GB, zero-valued guard bands.

implemented with an M:1 multiplexer: a $\log_2 M$-bit input signal selects a complex value out of the M prestored constants that form the constellation. In the implementation used for this work, Gray mapping and average power normalization are considered in the definition of the constellation point values.

After digital modulation, the *subcarrier mapping* module is responsible for mapping the A input active subcarriers to the central bins of an N-element array and zeroing the centre bin (the DC null subcarrier). The remaining $N - A - 1$ bins correspond to null subcarriers that serve as guard bands. As the IFFT DC bin is at index 0, an IFFT shift operation is performed on the N-element array. The resulting vector is then fed to the IFFT core. Here, it is assumed that the A active subcarriers include both data and pilot subcarriers and that the higher levels of the communication system provide them in their correct relative locations. The main modules required for *subcarrier mapping* are a double buffer and a control unit. The double buffer is implemented with a dual-port RAM, with each half storing N complex samples. This allows for simultaneous reading and writing of consecutive A-element arrays without any data conflicts: while one buffer is used for input (writing), the other is used for output (reading). The read/write access to the double buffer is managed by a control unit that receives and correctly maps the data to the correct IFFT input bin. The index mapping scheme implemented by the control unit combines subcarrier mapping and IFFT shifting.

The IFFT module implements a Cooley-Tukey Mixed-Radix algorithm using a pipelined single-delay feedback architecture as in [16]. The IFFT module has several processing stages which are comprised of shift registers, ROM memories, complex multipliers and arithmetic blocks (called "butterflies"). Information on the internal structures of Radix-2^2 and Radix-2 butterflies can be found in [17, 18]. Apart from processing elements, the IFFT module also includes blocks for input data reordering and bit-reversed reordering of intermediate results, which are performed with RAM-based double buffers.

The IFFT module produces time-domain OFDM symbols. The next module in the datapath is responsible for *cyclic prefix (CP) insertion*. It receives a data array of size N (corresponding to a time-domain OFDM symbol) and stores it in memory. Then, the module starts to read and output the last $L_{CP} + W$ memory positions. The cyclic prefix extension by W samples allows for the following weighted overlap and add (*WOLA*) operation. After outputting the last $L_{CP} + W$ memory positions, the CP insertion unit continues by reading and outputting the complete OFDM symbol from the beginning. Thus, the output of this module is an extended ODFM symbol with $N + L_{CP} + W$ complex samples. Its main hardware elements are an N-elements dual-port RAM and a unit for controlling write/read memory operations.

The final module performs the *WOLA* operation. It can be divided into two stages: first, OFDM symbols are multiplied by a window (*windowing*), and then the symbol's tail is overlapped and added with next symbol's head (*overlap-and-add*). The windowing operation is implemented using two multipliers (to handle the real and imaginary parts) and a ROM memory with prestored non-unitary raised-cosine window coefficients. In turn, the overlap-and-add operation is implemented with a finite-state machine (FSM) and arrays of registers to temporarily store each symbol's head and tail.

3.2 Baseband datapath for FBMC

The conceptual structure of the FBMC baseband modulator implemented for this work is shown in **Figure 2**. The *OQAM mapper* consists of two stages: first, the incoming data is QAM-modulated; then, the resulting in-phase and quadrature components are decoupled and alternately transmitted on successive subcarriers

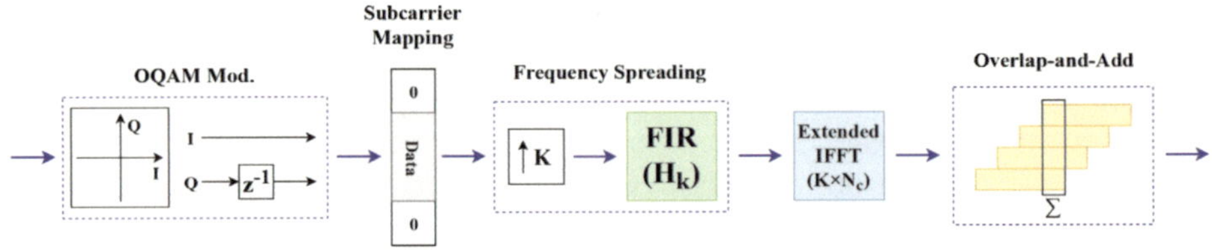

Figure 2.
Frequency spreading FBMC-OQAM baseband modulation.

and on successive transmitted symbols [19]. For instance, if the a symbol includes the in-phase (I) and quadrature (Q) components with the pattern *I,Q,I,Q,...*, the next symbol will use the pattern *Q,I,Q,I...*. The QAM mapper is implemented as for the OFDM modulator. The I/Q decoupling is efficiently performed with an FSM that alternately stores or outputs the I/Q components of a QAM symbol.

The following datapath modules are mainly characterized by parameters K and N_c. The *guard band insertion* module places the OQAM symbols in the central bins of an N_c-element array. The remaining subcarriers are zero and represent frequency guard bands. The operation of this module is similar to *subcarrier mapping* in OFDM modulation, except that that no DC null component is inserted.

The frequency spreading operation comprises *upsampling* by K and *FIR* filtering. The upsampler outputs $K - 1$ zero values between two incoming I/Q samples. It uses registers to store the input data and a counter to control the number of zero values at the output. For pulse shaping, a FIR filter architecture with a transpose structure was adopted because, unlike the direct FIR model, it does not require an extra input shift register, nor a tree of pipelined adders to achieve high throughput. The number of filter coefficients is odd $(2 \times K - 1)$, and their values are symmetric with a single-centre coefficient equal to one (**Table 4**). The multiplications by the centre coefficient can be ignored, as they do not affect the input value. However, the remaining coefficients imply non-trivial multiplications. The amount of non-trivial multiplications per FIR filter can be halved by exploiting the symmetry of the coefficients. As the sub-band signal is complex-valued, two FIR filters are required to separately filter the real and imaginary parts. The IFFT modules are the same as those used in the OFDM modulator.

The final operation is to *overlap-and-add* consecutive IFFT output stream blocks delayed by $N_c/2$ samples [19]. This operation uses an array of $2 \times K \times N_c$ elements as temporary storage; the first half stores the current FBMC symbol, and the second half accumulates IFFT output blocks. For each IFFT output block, the whole array is shifted by $N_c/2$ positions and then the IFFT output block is added to the second half of the array. A direct mapping of this approach to a hardware implementation would require the use of replicated memory structures to perform two read operations per clock cycle on the temporary array [21]. Instead, the overlap-and-add (OAA) module used was inspired by the architecture used in [22]. The main difference has to do with the fact that OQAM is not employed in [22] and, for the overlap-

K	H_0	$H_1 = H_{-1}$	$H_2 = H_{-2}$	$H_3 = H_{-3}$
2	1	$\sqrt{2}/2$	—	—
3	1	0.911438	0.411438	—
4	1	0.971960	$\sqrt{2}/2$	0.235147

Table 4.
Frequency domain prototype filter coefficients [20].

and-add operation, the consecutive IFFT output block streams are delayed by N_c. To continuously accumulate consecutive IFFT output blocks delayed by $N_c/2$, a feedback shift register of $(2 \times K - 1) \times N_c/2$ samples is used to align the previous IFFT block with the incoming IFFT block.

3.3 Baseband datapath for UFMC

UFMC, sometimes called Universal Filtered OFDM (UF-OFDM), is an OFDM-based waveform that attempts to reduce OOB emissions by time-domain filtering. The N subcarriers of each symbol are divided into B *physical resource blocks* (PRBs) of N/B subcarriers each. Usually, only part of the PRBs is used for transmission (*active PRBs*). For each active PRB, IFFT and bandpass L-order FIR filtering are performed. Instead of the CP, a zero-valued guard interval with length L is inserted after the IFFT. Frequency-shifted versions of the FIR filter are applied to all active PRBs, and, finally, the filtered sub-bands are superimposed to form an UFMC multicarrier symbol. Chebyshev filters are normally used for bandpass filtering in UFMC [23–25].

The classic UFMC modulation scheme [26] uses an N-point IFFT and FIR filters with complex coefficients for each active sub-band. To reduce this increased complexity, Knopp et al. [27] combine a smaller N'-point IFFT with N/N' upsampling. Moreover, the same real-coefficient FIR filter is used in all sub-bands, followed by frequency shifters implemented as multiplications by a complex exponential. **Figure 3** illustrates the datapath structure for the UFMC modulator considered in this work.

The UFMC modulator of this work has three processing branches, one to process each active PRB (B = 3). These branches share the same architecture and start with QAM mapping of the incoming data. The QAM mapper is equal to the one used in the OFDM and FBMC datapaths. The *subcarrier mapping* module maps the 12 PRB subcarriers to the central bins of an array with N' (64) elements and zeroes the remaining $N' - 12$ elements. It follows the same approach as subcarrier mapping in the FBMC modulator: a double buffer of $2 \times N'$ elements and read/write control engines.

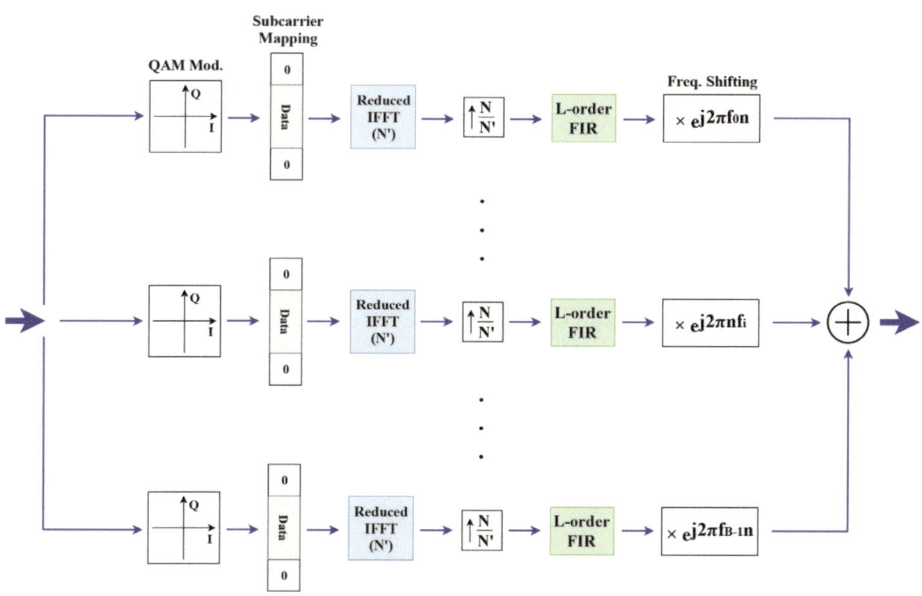

Figure 3.
Conceptual structure of the UFMC baseband modulator.

UFMC performs well for short-packet lengths and sporadic burst transmission [28, 29]. Moreover, the parallel sub-band processing in UFMC requires an IFFT core per branch. Therefore, instead of the high-performance pipelined IFFT architectures adopted for the OFDM and FBMC datapaths, low-resource memory-based FFT architectures are adopted in the UFMC modulator. The memory-based architecture adopted here is detailed in [30]. The *upsampler* architecture and operation is similar to the one used for frequency spreading in FBMC modulation. Here, the number of zeros between consecutive IFFT output samples is $(N/N') - 1$.

Dolph-Chebyshev FIR filters with a transpose structure are used for bandpass sub-band filtering. Again, the FIR coefficients are symmetric: there are an odd number of symmetric coefficients, and the centre coefficient is equal to one. However, the higher FIR order used in UFMC modulation requires further discussion. Considering an L-order FIR filter, $L - 1$ coefficients imply non-trivial multiplications that can be halved due to coefficient symmetry $(\frac{L-1}{2})$. As each processing branch requires two FIR filters—for the real and imaginary parts—there are $L - 1$ non-trivial multiplications per branch.

In Xilinx FPGAs, non-trivial multiplications can be efficiently performed by DSP blocks. These blocks are embedded into the logic fabric in a column arrangement. Cost-optimized devices have a smaller amount of DSP blocks, and their utilization should be carefully considered. For instance, the xc7z020 device has 220 DSP blocks. Considering the modes of operation from **Table 4**, the overall amount of non-trivial multiplications for FIR filtering $(3 \times (L - 1))$ is 108 for mode 1 and 216 for mode 2. This represents a high DSP utilization, and the sparse distribution of these types of blocks throughout the logic fabric degrades the scalability of the UFMC modulator. In addition, placement and routing of the design is more difficult and likely to affect the overall timing closure. To reduce DSP utilization, a multiplier-less architecture for FIR filters was adopted. The FIR coefficients are represented in Q1.5 format, using the Canonic Signed Digit (CSD) system with minimum non-zero bits. Then, non-trivial multiplications are substituted by shifters and adders. For example, the multiplication by 0.90625 can be implemented as:

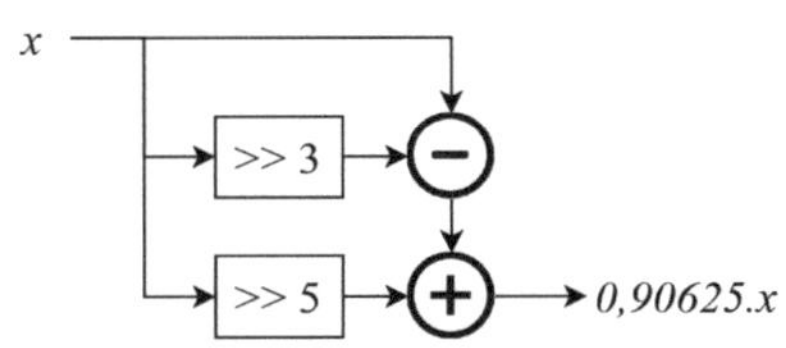

$$
\begin{aligned}
0.90625_{10} &= 0.11101_2 \\
&= 1.00\bar{1}01_{2,CSD} \\
&= (1 - 2^{-3} + 2^{-5})_{10}.
\end{aligned}
$$

This strategy eliminates the use of DSP blocks in FIR filters, but increases slice utilization. However, slices are the most numerous type of resource (13,300 slices in the xc7z020 device), making this approach well-suited for the present application.

After FIR filtering, each sub-band signal is shifted to the corresponding frequency band. The *frequency shift* module for each branch has a ROM memory to store the complex exponential values and a complex multiplier. Finally, the filtered sub-band responses are summed to create the UFMC symbol.

4. A dynamically reconfigurable baseband modulator for 5G communication

After the preceding overview of the architecture of high-performance baseband engines for three different waveforms, this section presents the architecture of a baseband processing engine that is *flexible*, *scalable*, resource and power *efficient* and *forward compatible*. Here, DPR and DFS are combined to produce a dynamically reconfigurable baseband processing architecture for multimode, multi-waveform coexistence and dynamic spectrum aggregation. To enable the full potential of 5G, carrier aggregation should also be possible across separated frequency bands [31] (*noncontiguous CA*). For noncontiguous CA, a *multidimensional* PHY layer (and, therefore, baseband architecture) is needed, even when data aggregation is not performed in the PHY layer, but in the media access control (MAC) communication layer instead [32]. In this context, multidimensional means that the PHY layer is an array of independent processing blocks, rather than a monolithic structure.

The baseband architecture presented in this chapter features three independent modulators, whose functionality and clock frequency can be dynamically reconfigured through DPR and DFS, respectively. This setup enables the processing of multiple component carriers with different waveforms and/or baseband parameters in noncontiguous CA schemes.

A prototype of the multidimensional baseband modulator was implemented on an Avnet Zedboard equipped with a Zynq xc7z020 device. The system top level combines features from the designs of the previous section and can be divided into three parts. The Zedboard's 512 MB DDR memory is used as a repository for the partial bitstreams used for DPR. The Zynq's ARM CPU act as the system management unit: it is responsible for triggering the reconfiguration of the multidimensional baseband modulator and setting up data transfers between the DDR memory and the modulators implemented in the programmable logic together with the infrastructure for DPR and DFS. **Figure 4** shows the top-level architecture.

The proposed architecture targets the communication scenario described in [33], which combines *multi-waveform coexistence* with *dynamic spectrum access*. In this scenario, 5G communications build on the pre-existing 4G infrastructure (*non-stand-alone 5G*): the primary 4G-LTE communications are OFDM-based, and the secondary 5G communications opportunistically exploit vacant spectrum resources through DSA, transmitting with different waveforms (OFDM, FBMC or UFMC). The basic unit for DPR is a complete baseband datapath, and each one is implemented in a reconfigurable partition. From the three RPs, RP_1 is exclusively used for primary OFDM-based transmission. The two remaining RPs can be used for primary or secondary transmission: RP_2 implements FBMC or OFDM transmission modes; RP_3 implements UFMC or OFDM transmission modes. For instance, if the primary transmission requires more capacity, the three RPs can be used to independently modulate different component carriers in a noncontiguous CA scheme. If the primary transmission is not so demanding, RP2 and RP3 can be used for secondary *multi-waveform* 5G transmission. **Figure 5** illustrates a potential multi-waveform coexistence scenario by showing the combined periodograms of the OFDM, FBMC and UFMC baseband signals obtained from the implemented modulator datapaths.

During system initialization, the ARM CPU manages the downloading of partial bitstreams and input data files from an SD card to the DDR memory. For the purpose of validating the baseband engines, the input data is retrieved from the DDR and sent to the baseband modulator(s), and the results are stored back to the DDR (and used for validating the implementation). Each RP has an associated DMA controller to accelerate the access to the DDR.

To achieve the specialization of computation at runtime, the configuration interface adopted is the ICAP. This high-bandwidth internal interface permits the FPGA to reconfigure itself. Xilinx sets the maximum ICAP bandwidth at 400, for a 100 clock frequency and 32-bit data width [34]. Nevertheless, the ICAP can be overclocked to further enhance the reconfiguration throughput [35]. In the present

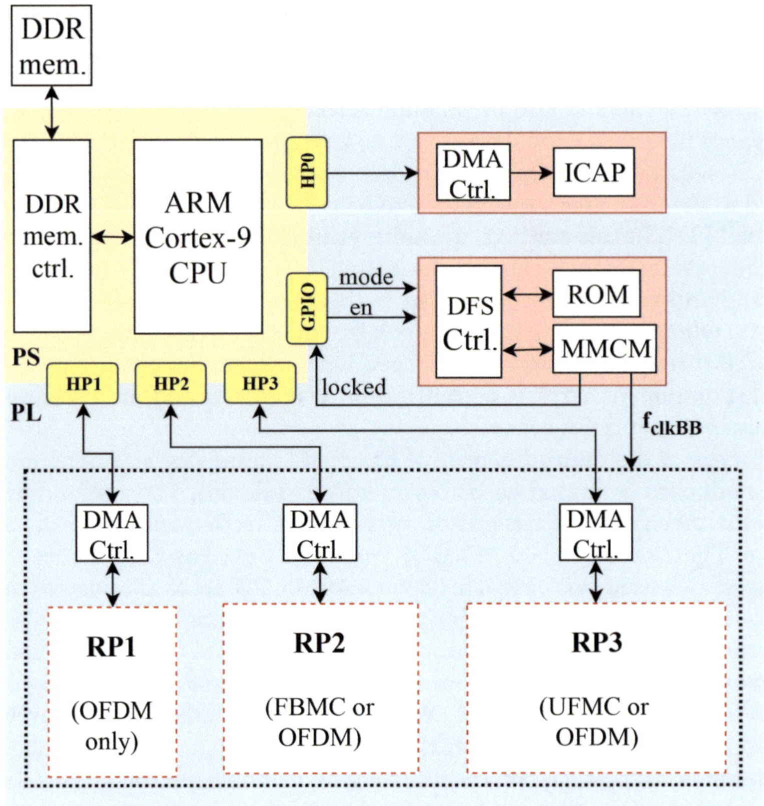

Figure 4.
Top-level architecture for the multidimensional and reconfigurable baseband modulator. HPx, high-performance ports; GPIO, general purpose I/O.

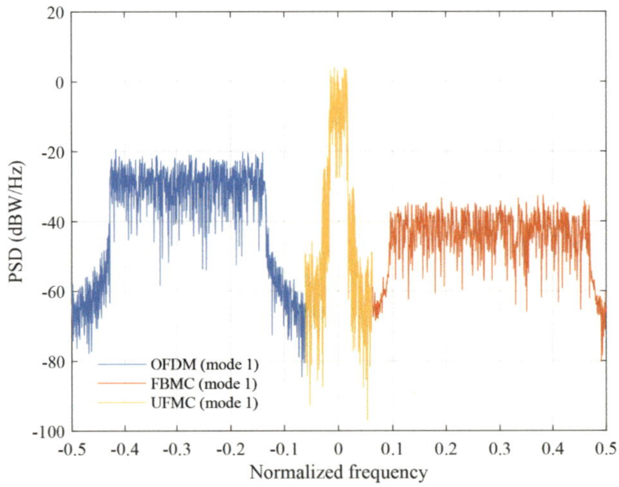

Figure 5.
Periodograms for OFDM, FBMC and UFMC baseband signals.

work, the ICAP is overclocked at 200 MHz. To take advantage of ICAP overclocking, a dedicated DMA controller is used to accelerate the transfer of partial bitstreams to the ICAP.

The implementation of DFS follows the reference design from [36]. This design considers an FSM that reads configuration parameters from a ROM and writes them to the clock management module available in the FPGA. To change the frequency of the output clocks, the input signal *en* must be enabled, and the desired mode of operation should be given through the *mode* port. The DFS controller is fed by a 100 reference input clock that is used to synthesize the clock signal used for baseband processing. Its frequency (f_{clkBB}) can be configured to one of four values: 16.7, 33.3, 66.7 and 100 MHz. All modulator datapaths can work at 100 MHz. The other values are based on the scaling of subcarrier spacing by 2^{μ} as in 5G New Radio systems [2], where μ is an integer that specifies the mode of operation. In this system, primary communications are based on the LTE OFDM numerologies (**Table 1**), where the subcarrier spacing (Δf) is 15 kHz. For OFDM mode 2 [cf. (**Table 1**)], the sampling frequency required is $N \times \Delta f = 15.36$ MHz. Scaling the subcarrier spacing by 2^{μ} with $\mu = \{1, 2\}$, results in sampling frequencies of $2^1 \times 15.36 = 30.72$ MHz and $2^2 \times 15.36 = 61.44$ MHz.

A general overview of the resource utilization of the prototype is presented in **Table 5**. The static part occupies around 32 and 5% of the slices and BRAMs, respectively. Apart from PS-PL interconnect cores and DMA controllers to accelerate the baseband modulators, the static part also implements the infrastructure for reconfiguration (DPR and DFS). The hardware required to implement DPR and DFS is below 2% of the available LUTs, FFs and BRAMs. The three RPs form the system's reconfigurable part and occupy 52.6, 64.3 and 72.7% of the available slices, BRAMs and DSPs, respectively. Overall, the resource utilization for the complete system implementation represents a considerable share of the resources available in the xc7z020 device: 84.3% of slices, 69.7% of BRAMs and 72.7% of DSPs.

The resource utilization of each modulator is presented in **Table 6**. The results lead to a key observation: the hardware virtualization achieved with the 7000 slices,

Resource	Available	Static part (total)	Reconfig. overhead		RP_1	RP_2	RP_3	All RPs
			DFS	DPR				
Slice	13,300	4210	24	424	1400	2400	3200	7000
LUT	53,200	10,700	75	938	5600	9600	12,800	28,000
FF	106,400	13,110	79	1292	11,200	19,200	25,600	56,000
BRAM	140	7.5	0	1.5	20	40	30	90
DSP	220	0	0	0	40	80	40	160

Table 5.
Post place-and-route resource utilization for the static and reconfigurable system parts.

Resource	Mode 1			Mode 2		
	OFDM	FBMC	UFMC	OFDM	FBMC	UFMC
Slice	1015	1575	2315	1126	2210	3100
LUT	2829	5103	8090	3400	7876	11,782
FF	2107	2307	6279	2170	2284	9912
BRAM	7	19	11.5	10.5	40	11.5
DSP	14	21	18	14	21	18

Device, xc7z020; $f_{clk} = 100$ *MHz.*

Table 6.
Post place-and-route resource utilization for each baseband modulator datapath.

90 BRAMs and 160 DSPs reserved by the three RPs allows the implementation of six baseband modulators, which would need 11,322 slices, 99.5 BRAMs and 106 DSPs in total. Adding these *virtualized* resources to the static resources exceeds the available xc7z020 slices by 17%. This is an unequivocal demonstration of the resource efficiency benefits that DPR brings to multimode baseband processors. An equivalent static multimode design could benefit from the reuse of common hardware blocks between different modulator datapaths (especially between OFDM and FBMC datapaths). However, implementing the multidimensional baseband modulator as a static multimode design would be challenging given the resource budget available on cost-optimized devices like the xc7z020. There are FPGA/SoC devices with larger area and logic density. However, using them would decrease the system's cost-effectiveness: an FPGA with a larger chip area is more expensive and likely to consume more power [37].

Considering the modes of operation shown in **Tables 1–3**, and that all 3 RPs are in use, the proposed design supports 32 combinations of baseband modulators: *2 RP_1 modes × 4 RP_2 modes × 4 RP_3 modes*. The use of DPR simplifies system upgrade with new modes of operation in order to extend the system's useful lifetime. The addition of modes of operation is not limited by the available resources on the FPGA device, but instead by the resources reserved by the RPs and the capacity to store partial bitstreams (512 MB DDR memory, in this case).

During the DPR design with the Xilinx Vivado EDA tool, the different system configurations are created from a design checkpoint that saves the floorplanning and routing of the system's static part, leaving the RPs as empty *black boxes*. New configurations can be created by designing new circuit configurations for these black boxes and generating the corresponding partial bitstreams. This design reusability makes the system adaptable and reduces the upgrade design time.

The dynamic power consumption for each modulator datapath and baseband clock frequency was estimated with the power analysis tool from Vivado 2015.2. The high-confidence estimates were performed using placed and routed netlists and accurate node activity files. The results are presented in **Table 7**. The UFMC modulator modes have a higher dynamic power consumption compared to FBMC and OFDM. This is mainly due to the higher resource usage and node activity of UFMC datapaths. The clock frequency adaptation allowed by DFS results in power savings that tend to be more evident for the most resource-demanding modes of operation (UFMC and FBMC). Compared to a design with baseband clock frequency fixed at 100 MHz, the clock frequency adaptation to:

- 66.7 MHz results in dynamic power savings between 39 mW (35% reduction in OFDM mode 1) and 82 mW (51% reduction in FBMC mode 2)

f_{clk}	**Mode 1**			**Mode 2**		
	OFDM	**FBMC**	**UFMC**	**OFDM**	**FBMC**	**UFMC**
100 MHz	113	148	180	123	161	233
66.7 MHz	74	84	119	78	79	155
33.3 MHz	34	25	60	33	28	77
16.7 MHz	14	8	30	10	10	39

Device, xc7z020; analysis tool, Vivado 2015.2; post place-and-route power analysis with high confidence level; node activity derived from post place-and-route simulation.

Table 7.
Dynamic power consumption estimates for the six implemented baseband modulator cores (in).

- 33.3 MHz results in dynamic power savings between 79 mW (70% reduction in OFDM mode 1) and 156 mW (67% reduction in UFMC mode 2)

- 16.7 MHz results in dynamic power savings between 99 mW (88% reduction in OFDM mode 1) and 194 mW (83% reduction in UFMC mode 2)

For the set of baseband clock frequencies defined, the DFS procedure took on average 47 μs to modify the clock frequency, a latency which is acceptable in 5G NR communications.

In the multidimensional baseband modulator, the area and amount of RP resources are higher than in the individual designs, resulting in larger bitstream sizes. However, the reconfiguration speed was increased through ICAP overclocking. **Table 8** quantifies the DPR latency and compressed bitstream size for the worst-case scenario in each RP. The largest RP (RP_3) takes up to 767 μs to be reconfigured, corresponding to the transfer of a 596 kB bitstream to the ICAP. In all DPR latency measurements, the reconfiguration throughput was at least 790 MB/s. This value is about 99% of the theoretical ICAP throughput, considering 32-bit transfers and overclocking at 200 MHz. In general, the DPR latency for each individual RP is below 1 ms, while the overall reconfiguration of the three RPs takes less than 2 ms. These latency values are within an acceptable range considering the control plane requirements from [38].

The ITU report [38] states that in critical, ultralow-latency scenarios, a *make-before-break* approach must be adopted to completely mitigate the control plane latency. In other words, the control plane latency must be *hidden* by setting up a new communication channel before breaking the current one. Under these circumstances, a high-priority communication can reserve a spare RP to seamlessly adapt the transmission mode. This scenario is exemplified in **Figure 6**. Let us assume that RP_1 is currently performing baseband modulation for an ultralow-latency communication. This transmission needs to be adapted from OFDM mode 1 to OFDM mode 2, without breaking the current communication link. RP_2 is currently unused and is reconfigured to OFDM mode 2 before baseband processing at RP_1 terminates. In this way, the baseband processing datapath can be modified without incurring any latency penalty due to DPR.

Characteristic	RP_1	RP_2	RP_3
DPR latency	400 μs	677 μs	767 μs
Partial bitstream size	309 kB	526 kB	596 kB

Table 8.
Measured DPR latency and size of compressed partial bitstreams for the worst-case scenarios.

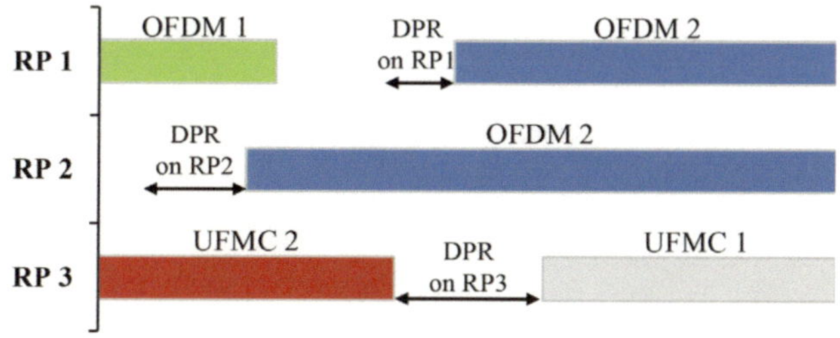

Figure 6.
Example of make-before-break approach to mitigate DPR latency.

5. Conclusion

This chapter presents a reconfigurable, multidimensional baseband modulator architecture suitable for multimode, multiple waveform coexistence and dynamic spectrum aggregation scenarios. The design combines the runtime specialization of computation and performance. By featuring three independent and reconfigurable baseband modulators, the architecture allows the processing of up to three component carriers using different waveforms (OFDM, FBMC and UFMC) and/or numerologies. The total reconfigurable area of the system covers more than half the available xc7z020 resources; the ICAP overclocking contributes to maintain the DPR latency low enough for the analyzed scenarios. In this design, the performance specialization through DFS resulted in dynamic power savings of up to 194 mW. Besides flexibility, scalability and forward compatibility, *cost-effectiveness* is perhaps the most relevant feature of this architecture. It is clearly demonstrated how the hardware virtualization through DPR enables implementations that exceed the hardware resources available on an FPGA device. This allows for system implementations on a small-form, cost-optimized devices with immediate cost and power consumption benefits and without compromising system functionality.

Acknowledgements

This work was financed by the ERDF (European Regional Development Fund) through the Operational Programme for Competitiveness and Internationalization (COMPETE) 2020 Programme within Project POCI-01-0145-FEDER-006961 and by the National Fund through a Ph.D. Grant (PD/BD/105860/2014) from the FCT (Fundação para a Ciência e a Tecnologia) (Portuguese Foundation for Science and Technology).

Author details

Mário Lopes Ferreira and João Canas Ferreira*
INESC TEC and University of Porto, Porto, Portugal

*Address all correspondence to: jcf@fe.up.pt

References

[1] ITU-R. IMT Vision—Framework and Overall Objectives of the Future Development of IMT for 2020 and beyond. ITU-R; 2015. ITU-R M.2083-0

[2] TS G. NR; NR and NG-RAN Overall Description; Stage 2 (Release 15); 2018. 38.300 V15.3.1. Available from: http://www.3gpp.org/DynaReport/38-series.htm

[3] Andrews JG, Buzzi S, Choi W, Hanly SV, Lozano A, Soong ACK, et al. What will 5G be? IEEE Journal on Selected Areas in Communications. 2014;**32**(6):1065-1082

[4] Luo FL, Zhang C. Signal Processing for 5G: Algorithms and Implementations. United Kingdom: John Wiley & Sons Ltd; 2016

[5] Jue G. Exploring 5G Coexistence Scenarios Using a Flexible Hardware/Software Testbed—Application Note; 2017

[6] Zhao Q, Sadler BM. A survey of dynamic Spectrum access. IEEE Signal Processing Magazine. 2007;**24**(3):79-89

[7] Akyildiz IF, Nie S, Lin SC, Chandrasekaran M. 5G roadmap: 10 key enabling technologies. Computer Networks. 2016;**106**:17-48

[8] Crockett LH, Elliot RA, Enderwitz MA, Stewart RW. The Zynq Book: Embedded Processing with the ARM Cortex-A9 on the Xilinx Zynq-7000 all Programmable SoC. Glasgow, United Kingdom: Strathclyde Academic Media; 2014

[9] Delahaye JP, Palicot J, Moy C, Leray P. Partial reconfiguration of FPGAs for dynamical reconfiguration of a software radio platform. In: 16th IST Mobile and Wireless Communications Summit. Budapest: IEEE; 2007. pp. 1-5. DOI: 10.1109/ISTMWC.2007.4299250

[10] Delorme J, Martin J, Nafkha A, Moy C, Clermidy F, Leray P, et al. A FPGA partial reconfiguration design approach for cognitive radio based on NoC architecture. In: 2008 Joint 6th International IEEE Northeast Workshop on Circuits and Systems and TAISA Conference. Montreal, QC: IEEE; 2008. pp. 355-358. DOI: 10.1109/NEWCAS.2008.4606394

[11] He K, Crockett L, Stewart R. Dynamic reconfiguration technologies based on FPGA in software defined radio system. Journal of Signal Processing Systems. 2011;**69**(1):75-85

[12] Vipin K, Fahmy SA. Mapping adaptive hardware systems with partial reconfiguration using CoPR for Zynq. In: 2015 NASA/ESA Conference on Adaptive Hardware and Systems (AHS). Montreal, QC: IEEE; 2015. pp. 1-8. DOI: 10.1109/AHS.2015.7231169

[13] Rihani MAF, Mroue M, Prévotet JC, Nouvel F, Mohanna Y. ARM-FPGA-based platform for reconfigurable wireless communication systems using partial reconfiguration. EURASIP Journal on Embedded Systems. 2017;**2017**(1):35

[14] Shreejith S, Banarjee B, Vipin K, Fahmy SA. Dynamic cognitive radios on the Xilinx Zynq Hybrid FPGA. In: Weichold M, Hamdi M, Shakir M, Abdallah M, Karagiannidis G, Ismail M, editors. Cognitive Radio Oriented Wireless Networks. CrownCom 2015. Lecture Notes of the Institute for Computer Sciences, Social Informatics and Telecommunications Engineering. Vol. 156. Cham: Springer; 2015. DOI: 10.1007/978-3-319-24540-9_35

[15] Pham TH, Fahmy SA, McLoughlin IV. An end-to-end multi-standard OFDM transceiver architecture using FPGA partial reconfiguration. IEEE Access. 2017;**5**:21002-21015

[16] Ferreira ML, Barahimi A, Ferreira JC. Reconfigurable FPGA-based FFT processor for cognitive radio applications. In: Proceedings of the Applied Reconfigurable Computing: 12th International Symposium, ARC 2016; March 22–24 March 2016; Mangaratiba, RJ, Brazil: Springer International Publishing; 2016. pp. 223-232

[17] He S, Torkelson M. A new approach to pipeline FFT processor. In: Proceedings of International Conference on Parallel Processing. Honolulu, HI, USA: IEEE; 1996. pp. 766-770. DOI: 10.1109/IPPS.1996.508145

[18] Löfgren J, Nilsson P. On hardware implementation of radix 3 and radix 5 FFT kernels for LTE systems. In: 2011, NORCHIP. Lund: IEEE; 2011. pp. 1-4. DOI: 10.1109/NORCHP.2011. 6126703

[19] Doré JB, Gerzaguet R, Cassiau N, Ktenas D. Waveform contenders for 5G: Description, analysis and comparison. Physical Communication. Elsevier; 2017;**24**:46-61. DOI: 10.1016/j. phycom.2017.05.004. ISSN: 1874-4907

[20] Bellanger M, Ruyet DL, Roviras D, Terr'e M, Nossek J, Baltar L, et al. FBMC physical layer: A primer. PHYDYAS Project; 2010

[21] Carvalho M. FPGA implementation of a baseband processor for FBMC transmission [MSc thesis]. Faculty of Engineering of the University of Porto; 2017

[22] Bellanger M. FS-FBMC: An alternative scheme for filter bank based multicarrier transmission. In: 2012 5th International Symposium on Communications, Control and Signal Processing. Rome: IEEE; 2012. pp. 1-4

[23] Wang X, Wild T, Schaich F, dos Santos AF. Universal filtered multi-carrier with leakage-based filter optimization. In: European Wireless 2014; 20th European Wireless Conference. Barcelona, Spain: VDE; 2014. pp. 1-5

[24] Jafri AR, Majid J, Zhang L, Imran MA, Najam-ul-Islam M. FPGA implementation of UFMC based baseband transmitter: Case study for LTE 10MHz channelization. Wireless Communications and Mobile Computing. Hindawi; 2018;**2018**:1-12. Article ID: 2139794. DOI: 10.1155/2018/ 2139794

[25] Nadal J, Nour CA, Baghdadi A. Flexible hardware platform for demonstrating new 5G waveform candidates. In: 2017 29th International Conference on Microelectronics (ICM). Beirut: IEEE; 2017. pp. 1-4. DOI: 10.1109/ICM.2017.8268851

[26] Vakilian V, Wild T, Schaich F, ten Brink S, Frigon J. Universal-filtered multi-carrier technique for wireless systems beyond LTE. In: 2013 IEEE Globecom Workshops (GC Wkshps). Atlanta, GA: IEEE; 2013. pp. 223-228. DOI: 10.1109/GLOCOMW.2013. 6824990

[27] Knopp R, Kaltenberger F, Vitiello C, Luise M. Universal filtered multicarrier for machine type communications in 5G. In: Proceedings of EUCNC 2016, European Conference on Networks and Communications; 2016. Available from: http://www.eurecom.fr/publication/ 4910. Unpublished material provided by EURECOM

[28] Schaich F, Wild T, Chen Y. Waveform contenders for 5G— Suitability for short packet and low latency transmissions. In: 2014 IEEE 79th Vehicular Technology Conference (VTC Spring). Seoul: IEEE; 2014. pp. 1-5. DOI: 10.1109/VTCSpring.2014. 7023145

[29] Parvez I, Rahmati A, Guvenc I, Sarwat AI, Dai H. A survey on low

latency towards 5G: RAN, core network and caching solutions. IEEE Communications Surveys Tutorials. 2018;**20**(4):3098-3130

[30] Lopes Ferreira M, Canas FJ. An FPGA-oriented baseband modulator architecture for 4G/5G communication scenarios. Electronics. 2019;**8**(1):1-19

[31] Bhushan N, Ji T, Koymen O, Smee J, Soriaga J, Subramanian S, et al. Industry perspective—5G air Interface system design principles. IEEE Wireless Communications. 2017;**24**(5):6-8

[32] Yuan G, Zhang X, Wang W, Yang Y. Carrier aggregation for LTE-advanced mobile communication systems. IEEE Communications Magazine. 2010;**48**(2): 88-93

[33] Kaltenberger F, Knopp R, Vitiello C, Danneberg M, Festag A. Experimental analysis of 5G candidate waveforms and their coexistence with 4G systems. In: XAPP888 - MMCM and PLL Dynamic Reconfiguration. 2015. Available from: http://www.eurecom.fr/fr/publication/4725/download/cm-publi-4725.pdf. Unpublished material provided by EURECOM

[34] UG909 - Vivado Design Suite User Guide: Partial Reconfiguration; 2015

[35] Claus C, Ahmed R, Altenried F, Stechele W. Towards rapid dynamic partial reconfiguration in video-based driver assistance systems. In: Sirisuk P, Morgan F, El-Ghazawi T, Amano H, editors. Applied Reconfigurable Computing: Architectures, Tools and Applications. Springer: Berlin Heidelberg; 2010. pp. 55-67

[36] Tatsukawa J. XAPP888 - MMCM and PLL Dynamic Reconfiguration; V1.7. Xilinx Inc.; April 2017

[37] Vipin K, Fahmy SA. FPGA dynamic and partial reconfiguration: A survey of architectures, methods, and applications. ACM Computing Surveys. 2018;**51**(4):1-39

[38] ITU-R. Minimum Requirements Related to Technical Performance for IMT-2020 Radio Interface(s). ITU-R; 2017. M.2410–0. Available from: https://www.itu.int/pub/R-REP-M.2410-2017

Chapter 4

An Efficient FPGA-Based Frequency Shifter for LTE/LTE-A Systems

Felipe A.P. de Figueiredo and Fabbryccio A.C.M. Cardoso

Abstract

The Physical Random Access Channel plays an important role in LTE and LTE-A systems. Through this channel, the user equipment aligns its uplink transmissions to the eNodeB's uplink and gains access to the network. One of the initial operations executed by the receiver at eNodeB side is the translation of the channel's signal back to base-band. This operation is a necessary step for preamble detection and can be executed through a time-domain frequency-shift operation. Therefore, in this paper, we present the hardware architecture and design details of an optimised and configurable FPGA-based time-domain frequency shifter. The proposed architecture is based on a customised Numerically Controlled Oscillator that is employed for creating complex exponential samples using only plain logical resources. The main advantage of the proposed architecture is that it completely removes the necessity of saving in memory a huge number of long complex exponentials by making use of a Look-Up Table and exploiting the quarter-wave symmetry of the basis waveform. The results demonstrate that the proposed architecture provides high Spurious Free Dynamic Range signals employing only a minimal number of FPGA resources. Additionally, the proposed architecture presents spur-suppression ranging from 62.13 to 153.58 dB without employing any correction.

Keywords: LTE, LTE-A, 4G, PRACH, NCO, time-domain frequency shift, FPGA

1. Introduction

Long Term Evolution (LTE) technology is the next big step forward in cellular services. It is a 3GPP-defined standard that is able to provide uplink speeds of up to 50 megabits per second (Mbps) and downlink speeds of up to 100 Mbps. This new technology delivers several technical benefits to cellular networks. Its bandwidth can be scaled from 1.25 MHz up to 20 MHz [1–4].

In order to make LTE a true fourth generation (4G) technology, it was enhanced to meet the IMT Advanced requirements issued by the International Telecommunication Union (ITU). The necessary improvements are specified in 3GPP Release 10 and also known as LTE Advanced (LTE-A). The LTE-A technology increases the peak data rates to 1 Gbit/s in the downlink and to 500 Mbit/s in the uplink. LTE-A has several new features such as MIMO extensions (up to 4×4 for UL and up to

8×8 for DL), carrier aggregation, improvement of the performance at cell edge by supporting enhanced intercell interference coordination (eICIC) and relay nodes (RN) and uplink access enhancements such as simultaneous data and control information (physical uplink shared channel (PUSCH) and physical uplink control channel (PUCCH)) transmissions and clustered single-carrier frequency-division multiple access (SC-FDMA) [3].

In LTE and LTE-A, uplink physical random access channel (PRACH) is used for initial access requests from the user equipment (UE) to the evolved base station (eNodeB) and to obtain time synchronisation [3, 4]. In case of a need to access the network, a UE requests access by transmitting a random access (RA) preamble through PRACH [5]. The RA preamble is then detected by the PRACH receiver at eNodeB side, which estimates both the ID of the transmitted preamble and the propagation delay between UE and eNodeB. Then, the UE is time-synchronised according to a time alignment (TA) value (derived from the propagation delay estimate) transmitted from the eNodeB before the uplink transmission [6].

PRACH transmission opportunity is set by higher layers [7] and determines the frequency-domain location of the random access preamble within the physical resource blocks (RB). In this way, at eNodeB side, a fundamental operation before any attempt to detect random access preambles takes place is the extraction of relevant preamble signals through a time-domain frequency shift operation. This operation translates the PRACH signal from the frequency-domain location set by higher layers back to baseband so that preamble detection can be totally carried out in baseband [4].

This paper is an extension of a previous conference paper [8]. Differently from [8], where we provided only some very superficial aspects of the proposed algorithm and architecture, the current paper presents a meticulous analysis on its design and implementation aspects. Therefore, the main contributions of the current paper are (i) the design of a low computational complexity time-domain frequency shifter algorithm and hardware architecture to be employed in the PRACH receiver at eNodeB side; (ii) a thorough analysis of design and implementation details; (iii) discussion of the computational complexity of the proposed architecture in terms of FPGA resource utilisation and speed; and (iv) careful analysis of the implementation results considering spur suppression, i.e. spurious-free dynamic range (SFDR), signal-to-noise ratio (SNR), probabilities of correct and error detection and average error between time-domain frequency shift operations carried out by a floating-point model, referred here as Golden Model (GM), and by the fixed-point FPGA implementation of the proposed architecture.

This paper contributes with a method and architecture optimised and tested for reduced complexity on a Xilinx Virtex-6 LX240T FPGA device. Results show that the architecture presents spur suppression better than 62 dB and when it is employed in the PRACH receiver, the probability of correct detection achieved by the receiver is greater than 99% at a SNR of −21 dB.

The remainder of the paper is organised as follows. In Section 2 we offer some background on the physical random access channel and its features. Section 3 presents an efficient algorithm for a time-domain frequency shifter. Section 4 gives important practical considerations on the implementation of the proposed algorithm as well as detailed description of the units composing the main - architecture. Test methodology, simulation and implementation results are then presented in Section 5. Finally, Section 6 provides some concluding remarks.

2. Physical random access channel

The PRACH is the physical channel that initiates the communication exchange with the eNodeB. Based on the sequences sent through this channel, the eNodeB is able to compute the time it takes for the signal to travel from the user equipment (UE) to it, identifying and correcting this time delay before establishing a data packet connection. In order to establish a connection with the eNodeB, the UE starts the random access procedure by transmitting the random access sequence (also known as preamble) through the PRACH. The PRACH preamble is made up of a cyclic prefix and a preamble part as presented in Table 5.7.1-1 of [7]. This preamble is orthogonal to other uplink user data to allow the eNodeB do differentiate each UE. The subcarrier spacing for the PRACH is 1.25 KHz for formats 0 to 3 and 7.5 KHz for format 4. See example in **Figure 1**. Formats 0 to 3 are used for frame structure type 1, i.e. frequency division duplexing (FDD), and Format 4 is used for the frame structure type 2, i.e. time division duplexing (TDD) only [3]. As will be discussed later, the PRACH can be positioned at different frequency locations, i.e. RBs, depending on a parameter configured by higher layers. **Figure 1** shows an example of a possible PRACH's frequency-domain location.

Prime-length Zadoff-Chu (ZC) sequences are employed as random access preambles in LTE and LTE-A systems due to their constant amplitude zero autocorrelation waveform (CAZAC) properties, i.e. all samples of a ZC sequence are located on the unit circle (unitary magnitude), and their autocorrelation values are equal to zero for all time-lags different from zero [9, 10]. These properties turn ZC sequences into very useful preambles for channel estimation, time synchronisation and improved performance of the detection of PRACH preambles [4]. ZC sequences transmitted through the PRACH channel present the form defined by Eq. (1) [7]:

$$z_u(n) = \exp\left(\frac{-j\pi u n(n+1)}{N_{ZC}}\right), 0 \le n \le N_{ZC} - 1, \tag{1}$$

where u is a positive integer known as ZC sequence index, n is the time index and N_{ZC} is the length of the ZC sequence, which for FDD systems is equal to 839 [7]. Random access preambles with zero correlation zones are defined from the uth root ZC sequence.

This sequence length, N_{ZC}, corresponds to approximately 69.92 physical uplink shared channel (PUSCH) subcarriers in each SC-FDMA symbol and offers a band

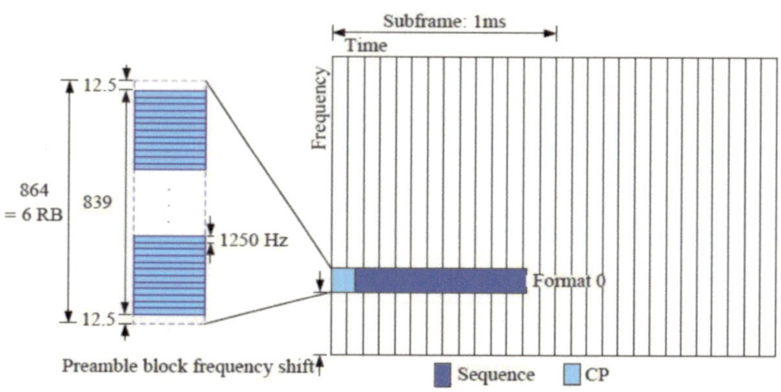

Figure 1.
Example of physical random access channel (PRACH) format 0.

protection of 72 − 69.92 = 2.08 PUSCH subcarriers, which corresponds to approximately one PUSCH subcarrier protection on each side of the preamble [7].

PUSCH subcarriers are spaced 15 KHz apart from each other.

The PRACH occupies a bandwidth of 1.08 MHz that is equivalent to six resource blocks (RB). Differently from other uplink channels, PRACH uses a subcarrier spacing of 1250 Hz for preamble formats 0 to 3 [7]. The ZC sequence is specifically positioned at the centre of the 1.08 MHz bandwidth, i.e. at the centre of the block of 864 available PRACH subcarriers, so that there is a guard band of 15.625 KHz on each side of the preamble, which corresponds to 12.5 null PRACH subcarriers. These guard bands are added to PRACH preamble edges in order to minimise interference from PUSCH. **Figure 1** depicts the PRACH preamble mapping according to what was just exposed.

The PRACH sequence, which for formats 0 and 1 is 800 us long, is created by cyclically shifting a root ZC sequence of prime-length N_{ZC}, defined as in Eq. (1). Random access preambles with zero correlation zones of length $N_{CS} - 1$ are generated by applying cyclic shifts to the uth root ZC sequence, according to Eq. (2):

$$x_{u,v}(n) = x_u((n + Cv) \bmod N_{ZC}), \tag{2}$$

where v is the sequence index and Cv is the cyclic shift applied to the root ZC sequence and calculated as $Cv = vN_{CS}$ for unrestricted sets [7]. The parameter N_{CS} gives the fixed length of the cyclic shift. All the possible values for these parameters are defined in [7].

2.1 PRACH receiver

In the literature there are two approaches for PRACH receivers, the full frequency domain and the hybrid time/frequency domain [4, 11]. Although the full frequency-domain approach provides the optimal detection performance, this approach uses considerably large size discrete Fourier transform (DFT), to be more

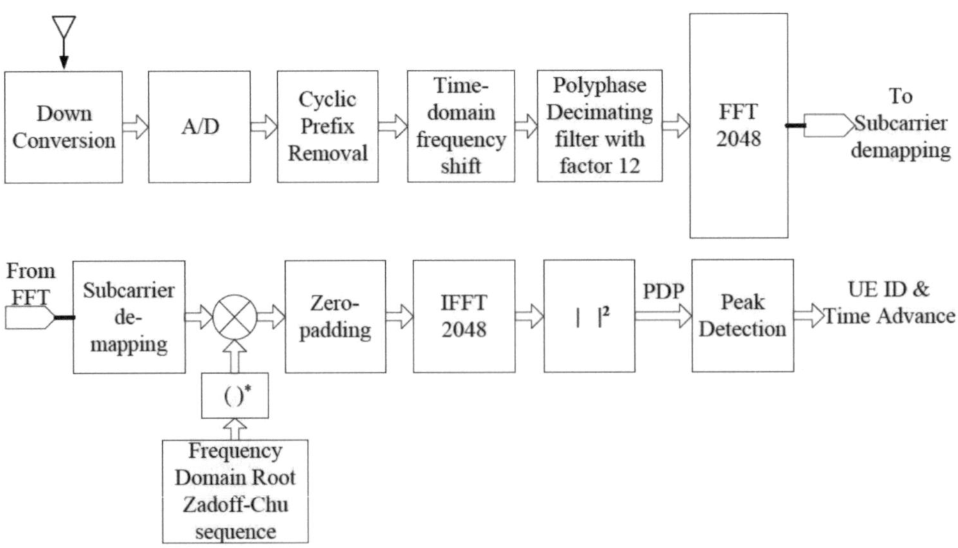

Figure 2.
Architecture of a hybrid time-/frequency-domain PRACH receiver.

precise a 24576-point DFT. On the other hand, the hybrid time-/frequency-domain approach uses FFT/IFFT blocks of the same size, i.e. 2048-point FFT when the decimation factor adopted is 12. Thus, the hybrid time/frequency domain substantially reduces the complexity of the hardware implementation. Therefore, in order to reduce the implementation complexity of the PRACH receiver, we adopt the hybrid time-/frequency-domain approach, which results in more practical implementations [4]. **Figure 2** depicts the PRACH receiver architecture implemented and being used in our L1 solution.

The received signal, i.e. possible random access preamble signal, is first preprocessed in time domain and then transformed into the frequency domain by an FFT block, multiplied by a Fourier transformed root Zadoff-Chu (ZC) sequence, and then the resulting sequence is searched for peaks above a predefined threshold which is calculated to produce a given probability of false alarm. **Figure 2** depicts the main components of the PRACH receiver at eNodeB side (for further details refer to [12]). The first block in **Figure 2** is the cyclic prefix remover, which discards all samples from the CP part of the preamble. Next, the PRACH pass-band signal is shifted to baseband by multiplying it with a complex exponential. In the sequence, the baseband signal is fed into a decimator block, which decimates the signal by a factor of 12; now instead of 24,576 samples in the case of format 0, we have only 2048 samples. The FFT block is responsible for transforming the SC-FDMA symbols from time domain into frequency domain. Next, the subcarrier demapping module extracts the RACH preamble sequence from the correct FFT bins. Then, the output of subcarrier demapping module is multiplied by the locally stored root ZC preamble, and then, the result of the multiplication is fed into the zero-padding module. Finally, the IFFT block is used to produce the cross-correlation between the root ZC sequence and the received preamble signal. All samples coming out of the IFFT block have their square modulus calculated producing what is known as power delay profile (PDP) samples. Finally, the preamble detection block employs the PDP samples to estimate the noise power, set a detection threshold and then decide whether a preamble is present or not. As an output of the detection process, this block reports to the MAC layer all detected preambles and its respective time advance (TA) estimates. For further information on this receiver architecture and detection algorithm, refer to [4, 12].

2.2 Preamble format

The PRACH preamble, illustrated in **Figure 3**, consists of three parts: a cyclic prefix (CP) with length T_{CP}, which is added to the preamble in order to effectively eliminate intersymbol interference (ISI) and a signature or sequence part of length

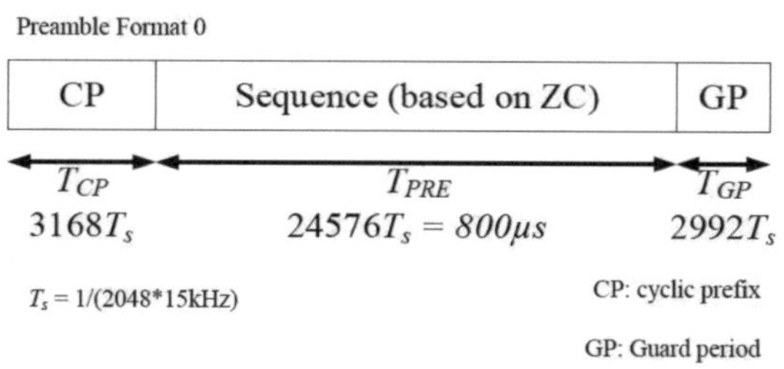

Figure 3.
Random access preamble format 0.

Preamble format	T_{CP}	T_{PRE}	T_{GP}
0	$3168.T_s$	$24576.T_s$	2976
1	$21024.T_s$	$24576.T_s$	15,840
2	$6240.T_s$	$2.24576.T_s$	6048
3	$21024.T_s$	$2.24576.T_s$	21,984

Table 1.
Random access preamble formats.

T_{PRE} and of a guard period T_{GP} which is an unused portion of time at the end of the preamble used for absorbing the propagation delay. The standard defines four different preamble formats for FDD operation [7]. Parameters T_{PRE}, T_{CP} and T_{GP} are set according to the chosen preamble format.

Figure 3 shows the parameter values for format 0, and the values for all formats are listed in **Table 1** where T_s is known as the standard time unit which is used throughout the LTE specification documents. It is defined as $T_s = 1/(15{,}000 \times 2048)$ seconds, which corresponds to a sampling rate of 30.72 MHz.

2.3 PRACH preamble signal

The PRACH preamble signal $s(t)$ can be defined as follows [7]:

$$s(t) = \beta_{PRACH} \sum_{k=0}^{N_{ZC}-1} \sum_{n=0}^{N_{ZC}-1} x_{u,v}(n).\exp\left[-\frac{j2\pi nk}{N_{ZC}}\right] \\ .\exp\left[j2\pi[k+\varphi+K(k_0+1/2)]\Delta f_{RA}(t-T_{CP})\right], \tag{3}$$

where $0 \le t < T_{PRE} + T_{CP}$, β_{PRACH} is an amplitude scaling factor and $k_0 = n_{PRB}^{RA} N_{SC}^{RB} - N_{RB}^{UL} N_{SC}^{RB}/2$. The location in the frequency domain is controlled by the parameter n_{PRB}^{RA} also known as $n_{PRB_{offset}}^{RA}$ (it is the input *frequency_offset_i* of the time-domain frequency shifter module) expressed as a resource block number configure by higher layers and fulfilling $0 \le n_{PRB}^{RA} \le N_{RB}^{UL} - 6$; this inequality is only valid for formats 0, 1, 2 and 3, i.e. FDD. The factor $K = \Delta f/\Delta f_{RA}$ accounts for the ratio of subcarrier spacing between the PUSCH and PRACH, and it is equal to 12 as Δf = 15 KHz and Δf_{RA} = 1250 Hz. The variable φ (equal to 7 for LTE FDD) defines a fixed offset determining the frequency-domain location of the random access preamble within the resource blocks. N_{RB}^{UL} is the uplink system bandwidth (in RBs), and N_{SC}^{BB} is the number of subcarriers per RB, i.e. 12.

By noticing that the inner summation is the DFT of $x_{u,v}(n)$ of length N_{ZC}, we can rewrite Eq. (3) in the following way:

$$s(t) = \beta_{PRACH} \sum_{k=0}^{N_{ZC}-1} X_{u,v}(k).\exp\left[j2\pi k \Delta f_{RA}(t-T_{CP})\right] \\ .\exp\left[j2\pi(\varphi+K(k_0+1/2))\Delta f_{RA}(t-T_{CP})\right]. \tag{4}$$

Again, by noticing that the first part of the summation in Eq. (4) is a time shift applied to the DFT of $x_{u,v}(n)$, we can rewrite that first part of the equation as follows by replacing t by Δt, which is referred in [7] as the standard time unit T_s, i.e. the sampling rate:

$$s(t) = \beta_{PRACH} \sum_{k=0}^{N_{ZC}-1} X_{u,v}(k) \\ .\exp\left[j2\pi k \Delta f_{RA} \Delta t(n - N_{CP})\right], \tag{5}$$

where N_{CP} is the number of samples corresponding to the CP interval as shown in **Figure 3** and $\Delta f_{RA}\Delta t$ = $1/N_{PRE}$ (where for formats 0 ad 1, N_{PRE} = 24,576).

Then rewriting Eq. (5), we have

$$s(t) = \beta_{PRACH} \sum_{k=0}^{N_{ZC}-1} X_{u,v}(k).\exp\left[\frac{j2\pi k(n - N_{CP})}{N_{PRE}}\right] \\ = \beta_{PRACH}.x'_{u,v}(n - N_{CP}). \tag{6}$$

Therefore as it can be easily seen, the result of the above equation is nothing more than the application of the DFT's time-shift theorem. It is also easy to see that this equation is the IDFT of $X_{u,v}(k)$ with length N_{PRE}. With that in mind, Eq. (4) can be rewritten as

$$s(t) = \beta_{PRACH}.x'_{u,v}(n - N_{CP}) \\ .\exp\left[\frac{j2\pi(\varphi + K(k_0 + 1/2))(n - N_{CP})}{N_{PRE}}\right]. \tag{7}$$

Eq. (4) can be reorganised in the following way:

$$s(t) = \beta^{PRACH.\exp}\left[\frac{-j2\pi N_{CP}(\varphi + K(k_0 + 1/2))}{N_{PRE}}\right] \\ .\left\{x'_{u,v}(n - N_{CP}).\exp\left[\frac{j2\pi n(\varphi + K(k_0 + 1/2))}{N_{PRE}}\right]\right\}. \tag{8}$$

The part of Eq. (8) between curly braces represents a circular frequency shift u, u applied to $x'_{u,v}$ $(n - N_{CP})$, i.e.

$$x'_{u,v}(n - N_{CP}).\exp\left[\frac{j2\pi nm}{N_{PRE}}\right] \overset{DFT}{\leftrightarrow} x_{u,v}(K - m), \tag{9}$$

where m is the frequency shift applied to the PRACH signal before it is transmitted and it is given by the following equation

$$m = \varphi + K(k_0 + 1/2). \tag{10}$$

Once we are only dealing with FDD, Eq. (10) can be further simplified as

$$m = 13 + 144n_{PRB}^{RA} - 72N_{RB}^{UL}. \tag{11}$$

Therefore, at the PRACH receiver side, after removing CP and GP, the preamble is still shifted in frequency domain by an offset factor given by m. For further processing, it is necessary to convert the shifted preamble into baseband. This conversion is performed by the time-domain frequency shift module (see **Figure 2**), which multiplies the received preamble by the conjugate of the complex exponential term given in Eq. (9).

3. Efficient algorithm of a time-domain frequency shifter

In this section, we present an efficient algorithm used to apply frequency-domain shifts to random access preamble signals in time domain (i.e. without the need to convert them to the frequency domain) through the use of a customised numerically controlled oscillator (NCO) and a complex multiplier. We also discuss the advantages presented by the proposed algorithm.

3.1 Numerically controlled oscillator

Numerically controlled oscillators (NCO) are important components in many digital communication systems. They are generally employed in quadrature synthesisers, which are used for constructing digital down- and upconverters and demodulators and here for time-domain frequency shifters. A very common method for creating digital complex or real valued sinusoid signals uses a lookup table (LUT) approach [13]. In this approach, a LUT saves into memory digital samples of a sinusoid signal.

A digital integrator is then employed to compute the correct phase arguments, which are mapped by the LUT to the desired output sinusoid samples. The integrator computes a phase slope that is mapped to a sinusoid (possibly complex) by the LUT. This value is presented to the address port of the LUT that performs the mapping from phase space to time [14].

A LUT usually saves into memory uniformly spaced samples of sine and cosine waveforms. This set of samples comprises a single cycle of a prototype complex sinusoid waveform with length $N = 2^{B_{\Theta(n)}}$ and consists of specific values of the argument $\Theta(n)$ of sinusoid waveform, as defined by Eq. (12).

$$\Theta(n) = n\frac{2\pi}{N}, \tag{12}$$

where n is the index of the time sample and $B_{\Theta(n)}$ is the number of bits employed in the phase accumulator which is calculated as shown in Eq. (13):

$$B_{\Theta(n)} = \log_2\left\lceil\frac{f_{clk}}{\Delta f}\right\rceil, \tag{13}$$

where $\lceil\cdot\rceil$ denotes the ceiling operator, f_{clk} is the system clock frequency and Δf is the frequency resolution of the NCO. The frequency resolution, Δf, of the NCO is a function of f_{clk} and $B_{\Theta(n)}$. Then Δf can be determined using the following equation:

$$\Delta f = \frac{f_{clk}}{2^{B_{\Theta(n)}}} = \frac{f_{clk}}{N}. \tag{14}$$

The output frequency, f_{out}, of the NCO waveform is a function of f_{clk}, $B_{\Theta(n)}$ and the phase increment value $\Delta\theta$. That is, $f_{out} = f(f_{clk}, B_{\Theta(n)}, \Delta\theta)$ which is given in Hertz and is defined in Eq. (15). The phase increment, $\Delta\theta$, is an unsigned value which defines the NCO output frequency:

$$f_{out} = \frac{f_{clk}\Delta\theta}{2^{B_{\Theta(n)}}} = \frac{f_{clk}\Delta\theta}{N}. \tag{15}$$

The accuracy of a signal sequence formed by reading samples of a sinusoid signal from a LUT is influenced by both the amplitude and the phase of the quantization process. The width and length of the LUT memory directly impact the resolution of both the signal's amplitude and phase angle. These resolution limits correspond to time base jitter and amplitude quantization of the signal, respectively. Additionally, these resolution limits add a white broadband noise floor and spectral modulation lines to the spectrum of the generated signal sequence [15].

Quarter-wave symmetry in the basis waveform can be exploited to construct an NCO that uses shortened tables. We will discuss this approach next.

3.2 Iterative time-domain frequency shift algorithm

The optimised algorithm representing the time-domain frequency shift operation is presented in Algorithm 1. It depicts the data processing executed by each one of the units in **Figure 4**.

At first, during eNodeB's initialisation, the parameters *offset* and *bandwidth* (*bw*) are sent by higher layers to the PHY which in turn feeds them into the time-domain frequency shifter module so that the discrete frequency shift calculator unit is able to calculate the actual frequency shift to be applied to the received PRACH signal. Whenever a subframe in which random access preamble transmissions are allowed happens (it is set according to Table 5.7.1-2 in [7]) and after CP is removed, the customised NCO unit generates a complex exponential signal with frequency set earlier by the discrete frequency shift calculator unit and multiplies it sample by sample with the incoming PRACH complex signal samples; note that it is a complex multiplication once both are complex signals.

The procedure inputs are *re_ad*, *im_ad*, *offset*, *bw* and *cos table* where *re_ad* and *im_ad* are the already CP removed quadrature samples coming from the analog to digital converter (ADC), *offset* and *bw* are the configuration parameters coming from higher layers and used to calculate the frequency shift necessary to translate the pass-band preamble signal back to baseband and *cos_table* is the LUT containing the samples of a sinusoid used to generate the complex exponential signal. The angle mapper is the main part of the customised NCO algorithm shown in Algorithm 1 once it maps *theta* into a value of a 1/4-length cosine table.

In the light of what was presented in the previous section, we now discuss Algorithm 1. The first part of the algorithm is responsible for calculating the discrete frequency of the complex exponential signal that the NCO must generate in order to

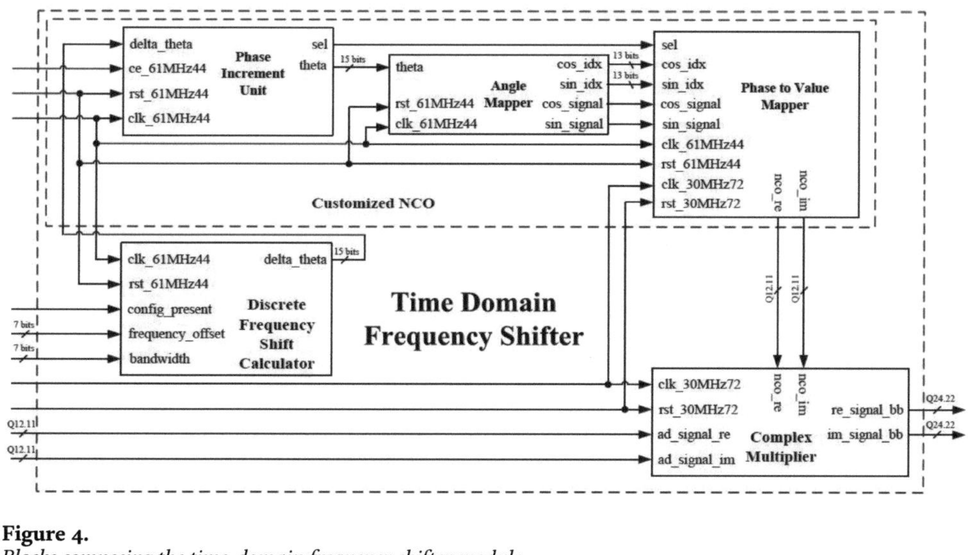

Figure 4.
Blocks composing the time-domain frequency shifter module.

shift the received pass-band preamble signal to baseband. The discrete frequency shift, m, is calculated as shown in Eq. (11). By remembering that $\Delta f_{RA}\Delta t = 1/N_{PRE}$, we can then rewrite the exponential part of Eq. (9) as

$$\exp\left[j2\pi\left(m\Delta f_{RA}\right)t\right]. \tag{16}$$

By analysing the equation above, it is noticeable that all frequency shifts are integer multiples of Δf_{RA}, and in this way we state that the output frequency of the NCO must be $f_{out} = m\Delta f_{RA}$. Before proceeding we must define the values for some parameters presented in the previous section. The frequency resolution $\Delta f = \Delta f_{RA} =$ 1250 Hz, the system clock frequency $f_{clk} = 1/T_s = 30.72$ MHz, the length, $N = 2^{B_{\Theta(n)}}$, of the single cycle of the basis complex waveform is made equal to 24,576 samples. The cycle length can be expressed as $N = f_{clk}/\Delta f_{RA}$.

Algorithm 1. Time-domain frequency shifter algorithm

```
procedure TDFreqShifter(re_ad, im_ad, offset, bw, cos_table)

    ▷ ********* Discrete Frequency Shift Calculator *********
    m = 13 + 144 * offset − 72 * bw;
    ▷ —- Frequency Control Word (FCW) —-
    delta_theta = m;
    if m < 0 then
        delta_theta = N + m;
    end if

     ▷ ************* customised NCO Algorithm ************
     theta = 0;
     for i ← 0 to N − 1 do
         ▷ —- Angle Mapper —-
         cos_signal = 1;
         sin_signal = 1;
         if theta > 3 * N/4 then
             cos_idx = (N − theta);
         else
             if theta > N/2 then
                 cos_idx = (theta − (N/2));
                 cos_signal = −1;
             else
                 if theta > N/4 then
                     cos_idx = ((N/2) − theta);
                     cos_signal = −1;
                     sin_signal = −1;
                 else
                     cos_idx = theta;
                     sin_signal = −1;
                 end if
             end if
         end if

         sin_idx = (N/4) – cos_idx;

         if cos_idx == N/4 then
```

```
            cos_idx = ((N/4) − 1);
        end if
        if sin_idx == N/4 then
            sin_idx = ((N/4) − 1);
        end if

        ▷—- Phase to Value Mapping (LUT) —-
        re_nco(i) = cos_signal ∗ cos_table(cos_idx);
        im_nco(i) = sin_signal ∗ cos_table(sin_idx);

        ▷—- Phase Increment (Phase Accumulator) —-
        theta = (theta + delta theta);
        if theta >= N then
            theta = (theta − N);
        end if

        ▷*************** Complex Multiplier ***************
        re_bb(i) = re_ad(i) ∗ re_nco(i) – im_ad(i) ∗ im_nco(i);
        im_bb(i) = re_ad(i) ∗ im_nco(i) + im_ad(i) ∗ re_nco(i);
    end for
    return re_bb, im_bb
end procedure
```

Then, using the aforementioned definitions and rewriting Eq. (15) letting $\Delta\theta$ in evidence, we have

$$\Delta\theta = \frac{f_{out}N}{f_{clk}} = \frac{m\Delta f_{RA}N}{f_{clk}} = m\frac{m\Delta f_{RA}}{f_{clk}}\frac{f_{clk}}{\Delta f_{RA}} = m. \tag{17}$$

In case m is negative, it is necessary to calculate its module in relation to N_{PRE} before feeding it into the customised NCO. Note in Algorithm 1 that the module operation is simply done by adding N_{PRE} to the negative value of m.

In a traditional NCO algorithm, i.e. one that adopts full-period waveforms, there would be two main parts, namely, phase accumulator and LUT. In its simplest form, there would be two LUTs storing samples of a cosine and a sine wave. However, this approach generally results very large tables, which sometimes are impractical. Therefore, for a practical implementation with reduced tables, the proposed algorithm employs only one LUT exploiting quarter-wave symmetry in the basis waveform and the constant phase offset ($pi/2$) between sine and cosine signals. In this approach we use one LUT with $N/4$ samples. However, when exploiting quarter-wave symmetry, the mapping from phase space to time is not direct as in the traditional NCO algorithm.

In order to exploit quarter-wave symmetry, an algorithm is needed to map the angle values (phase space), θ, output by the phase accumulator into valid positions of a shortened LUT containing the samples of a cosine signal. This task is performed by the angle mapper part of Algorithm 1. The angle mapper maps angle values in the second, third and fourth quadrants into the first one and tracks the signals that must be applied to cosine and sine values. As can be seen in Algorithm 1, the indices for generating the cosine signal are calculated first, and then a $N/4$-phase offset, which is equivalent to a $pi/2$ offset, is applied to it in order to generate the sine indexes. In order to store values ranging from 1 to 0, i.e. the first quadrant of a cosine signal, $(N/4) + 1$ samples would be necessary where the last one is zero.

The zero value can be mapped to the value stored at the $N/4$-th position with minimal degradation on SFDR performance. Therefore, as can be seen in the algorithm, when either sine or cosine indexes are equal to $N/4$, their values are changed to $(N/4) - 1$, which is the closest value to zero.

As the LUT only stores samples from the first quadrant of a cosine signal, i.e. only positive values, the phase to value mapper part of the algorithm must apply the correct signals (provided by the angle mapper) to the LUT's output.

The phase increment (also referred as phase accumulator) part of the algorithm acts as an integrator. It calculates at each iteration of the algorithm a new phase value, θ, by using the phase increment $\Delta\theta$ value provided by the discrete frequency shift calculator part. Once the angle mapper algorithm can only map values ranging from 0 to $N - 1$ into the range 0 to $(N/4) - 1$, it is necessary to apply a module operation in case the resulting phase value is equal or greater than N.

The module operation is easily performed by subtracting N from the phase value. The last part of the algorithm multiplies sample by sample the generated complex exponential signal by the ADC signal. That complex multiplication operation then translates the pass-band PRACH signal into baseband.

3.3 Advantage of the proposed algorithm

The main advantage of the proposed algorithm is the memory savings attained by the use of a 1/4-length cosine table instead of storing in RAM each one of the $2N$ possible samples of a complex exponential signal. **Table 2** shows the memory utilisation for some data widths if we were to store all the 2N samples (sine and cosine waves) corresponding to one complete period of the basis complex exponential waveform necessary to translate the received PRACH signal into baseband. In Xilinx FPGAs, a block RAM (BRAM) is a dedicated, i.e. they cannot be used for anything, but RAM, a two-port memory containing several kilobits of RAM. A FPGA contains a limited number of these blocks. The configuration logical blocks (CLB) in most of Xilinx FPGA contain a small RAM. They are called distributed (LUT) RAM because they are distributed throughout the FPGA once they are part of a CLB. This kind of RAM can normally store only a dozen bits. A reasonable rule of thumb when designing with FPGAs is that if you need a lot of RAM, as is the case here, you should use BRAMs; otherwise, the FPGA resources will be eaten up implementing the RAM in distributed RAM. It is important to say that a Virtex-6 FPGA device has only 416 36 kb BRAMs which makes it a very precious resource when implementing a large project as an L1 PHY, and in this way its usage must be taken into account during planning and development. One alternative to decrease the number of occupied BRAMs is the exploitation of quarter-wave symmetry in the basis waveform. This alternative results in a customised NCO that employs a shortened LUT, as can be seen in **Table 3**.

The fourth column in **Table 3** shows the reduction of used BRAMs when exploiting quarter-wave symmetry. As can be noticed, this approach results in a

Data width	Size in kb	No. of 36 kb BRAMs
8	384	11
12	576	16
16	768	22
24	1152	32

Table 2.
Memory utilisation when storing the full period of the complex exponential.

Data width	Size in kb	No. of 36 kb BRAMs	Reduction in %
8	48	2	81.8
12	72	2	87.5
16	96	3	86.4
24	144	4	87.5

Table 3.
Memory utilisation when employing quarter-wave symmetry.

more area efficient implementation because the memory requirements are minimised, i.e. fewer FPGA BRAMs are required. Therefore this approach saves on valuable chip area and also reduces power consumption.

4. Implementation details

This section presents some discussions on implementation details of the proposed architecture. It is suitable for implementation on devices that employ hardware description language (HDL) as part of its design process such as field-programmable gate arrays (FPGA) and application-specific integrated circuits (ASIC). **Figure 4** shows the hardware architecture of the time-domain frequency shifter module. The architecture employs only one LUT and exploits quarter-wave symmetry for shortened tables.

The proposed architecture works with two different system clocks. The first clock, of 30.72 MHz, is used by the ADC unit and therefore dictates the rate the complex multiplication is performed. The complex multiplier and phase to value mapper modules are the only two modules running at this clock rate. The second clock rate employed in the system is 61.44 MHz and is used so that two samples of the complex exponential waveform can be read from the same LUT memory during the period of one sample arriving from the ADC module. This dual system clock scheme drastically reduces the amount of RAM memory necessary for the system to be implemented. All modules composing the proposed architecture run at this clock rate, the complex multiplier module being the only exception.

The two input parameters, *bandwidth* and *frequency_offset*, are fed into the module only when the eNodeB is being initialised. They can be considered static parameters of eNodeB. The input signal *config_present* is asserted by higher layers during the initialisation process to inform when the input parameter values are valid.

The proposed architecture has to be informed through the *ce_61MHz*44 signal when the sequence section of the RACH preamble starts so that it can be multiplied by the local complex exponential, which is generated by the customised NCO module. The signal *ce_61MHz*44 is set by the PRACH receiver module after it removes (i.e. discards) the CP portion of the RACH preamble and has to stay in high state level for the whole duration of the RACH sequence portion, e.g. in the case of format 0, the signal *ce_61MHz*44 must remain in high state for 24,576 30.72 MHz clock cycles, i.e. $2 \times 24{,}576$, when considering the 61.44 MHz system clock rate. It is important to highlight that some latency is expected once all modules have registered outputs, and therefore, this latency has to be taken into consideration by the PRACH receiver module when setting the enable signal of the chip.

Another important characteristic of the proposed architecture is that it only employs a total of four multipliers that are used by the complex multiplier module. Moreover, the proposed architecture only uses plain add and bit-shift operations to compute the value of trigonometric functions such as complex exponential sequences, which turns it into a highly efficient hardware architecture in terms of logical resource consumption.

Additionally, the proposed frequency shift architecture can have its inputs and outputs entirely configured, i.e. the width of input and output signals can be set to one of the following choices: 8, 12, 16 and 24 bits. In the case of our actual implementation, we employ an ADC with an output of 12 bits for each one of the quadrature components, i.e. in-phase (I) and quadrature (Q) components, and it has a fixed-point representation (Q-format) of Q12.11, i.e. 1 bit for the integer part and 11 bits for the fractional part. The I and Q components computed by the customised NCO module present the same fixed-point representation of the input of the frequency shifter module. In relation to this particular point, after the complex multiplication between the NCO and the ADC quadrature samples, which requires the multiplication and subsequent addition of samples, the fixed-point representation of the modules' output is equal to Q25.22. Since the maximum possible value generated by the complex multiplication operation is 2, the integer part only needs 2 bits instead of 3 bits. Therefore, depending on the selected width configuration, the fixed-point representation of the complex signal output by the module can be configured to Q8.6, Q12.10, Q16.14 and Q24.22.

4.1 Discrete frequency shift calculator unit

Figure 5 depicts the proposed architecture for the discrete frequency shift calculator module. This module is employed to compute the frequency shift, m, that must be applied to the received PRACH signal sequence in order to translate it into a baseband signal, i.e. a signal centred around 0 Hz. Therefore, in order to compute such frequency shifts, the module implements Eq. (11). All multiplications involved here are executed by bit-shifting the input values N_{RB}^{UL} and n_{PRB}^{RA} by the constant values -72 and 144, respectively, and then adding the result to the constant value 13. Before sending the value of m to the customised NCO module, it is necessary to verify whether the resulting value is negative or not; if it is negative, then the constant value N has to be summed to the result value, which turns m into a positive value. It is done due to the fact that the phase increment module, which composes the customised NCO module, only expects positive input values. As defined by Eq. (17), the discrete frequency shift, m, is equal to $\Delta\theta$, which is the necessary input value for the customised NCO module to operate properly.

4.2 Customised numerically controlled oscillator

The customised numerically controlled module is composed of three blocks, namely, phase increment, angle mapper and phase to value mapper, as can be seen

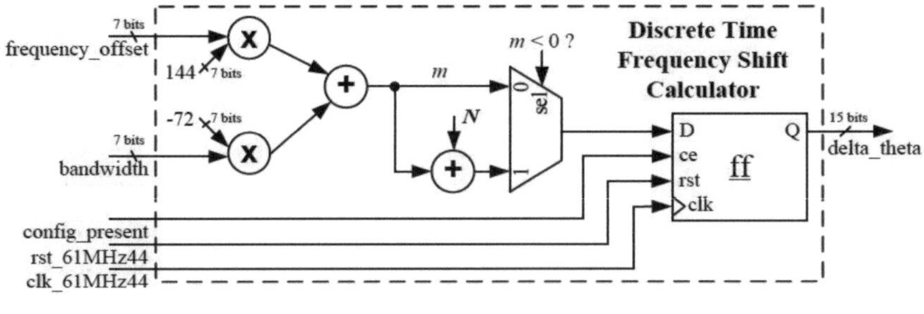

Figure 5.
Architecture of the discrete frequency shift calculator unit.

in **Figure 4**. The first two modules run at a system clock of 61.44 MHz, and the last one runs at 30.72 and 61.44 MHz since it is the module in charge of reading both quadrature components, I and Q, from the LUT memory inside one period of the 30.72 MHz system clock rate.

Figure 6 depicts the proposed architecture of the phase increment module. This module is the implementation of a digital integrator, which computes the phase argument, θ, sent to the angle mapper module. At each iteration, $\Delta\theta$ is added to θ, which starts from a value equal to 0. If θ results in a value that is greater than $N + 1$, then the constant value $-N$ is added to it so that its value remains less than N and, therefore, it can be correctly mapped into a valid phase argument value. The procedure we have just described is nothing but the direct implementation of the module operation, mod(θ, N). This module only produces a valid θ value when the selection signal, *sel*, is set to high level. The *sel* signal is produced by a system clock divisor that divides the 61.44 MHz system clock by 2, i.e. the module produces a valid output value at a clock rate of 30.72 MHz. The *sel* signal is also generated by the module in order to feed the phase to value mapper module. As Eq. (13) is equal to N and it is not equal to a power of 2, the data width of the phase increment module, $B_{\Theta(n)}$, is ceiled, then resulting in a value with 15 bits.

Figure 7 shows the proposed architecture for the angle mapper module. This module is in charge of translating the phase argument value, θ, which can vary in the range between 0 and $2 * \pi$ (i.e. from 0 to N) into a phase argument value inside the first quadrant of the circle, i.e. a value in the range between 0 and *pi*/2 (i.e. from 0 and $N/4$). The output value of this module is the index of cosine waveform, which is employed as an address value to access one of the $N/4$ values saved in the LUT memory. In order to compute the index of the sine waveform, a constant phase offset value equal to *pi*/2, i.e. $N/4$, has to be applied to the cosine index value. Since the value corresponding to $\cos(\pi/2)$, i.e. 0, is not saved into the LUT memory, whenever either cosine or sine index values are equal to $N/4$, then their index values are modified to $(N/4) - 1$, which is the closest index value to 0.

The module is also responsible for keeping track of the signals (i.e. +/−) that must be applied to the I and Q values at the output of the phase to value mapper module. These signals translate the phase argument value, which are represented by the sine and cosine index values, back to its original quadrant of the disc. At the input of this module, the phase argument, θ, presents a databus width of 15 bits so that it is able to access N samples stored in the LUT memory, i.e. all the four

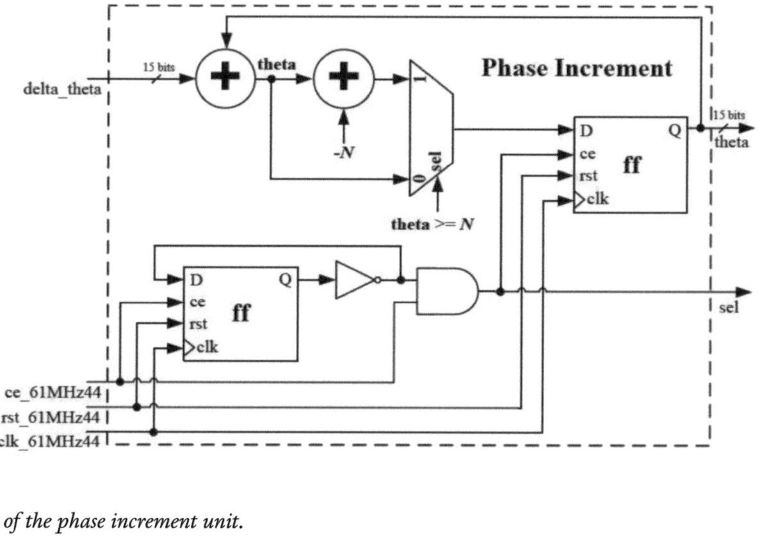

Figure 6.
Architecture of the phase increment unit.

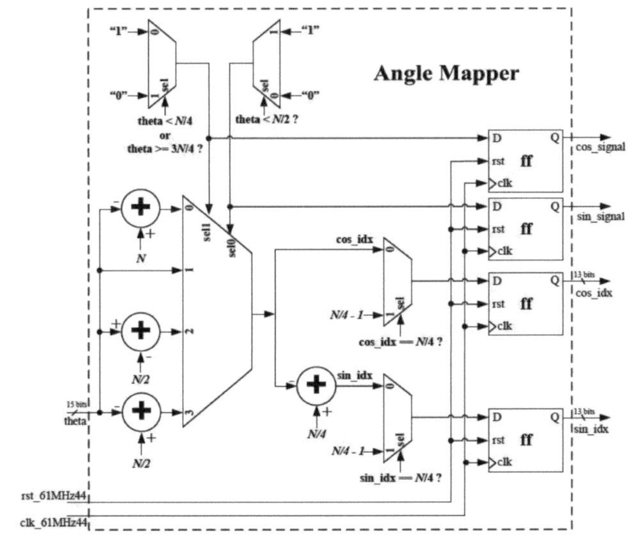

Figure 7.
Architecture of the angle mapper unit.

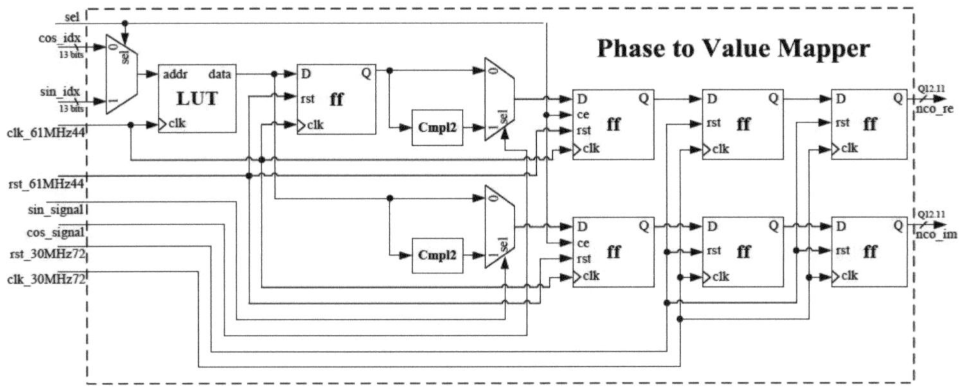

Figure 8.
Architecture of the phase to value mapper unit.

quadrants of the disc. Therefore, since the angle mapper module converts θ to the first quadrant of the disc, its databus width can be decreased to 13 bits, which is the number of bits used to access the $N/4$ samples of the first quadrant (i.e. quarter-wave symmetry) and that are saved in the LUT memory. This module runs at a clock rate of 61.44 MHz and outputs two new phase argument indexes, for the sine and cosine waveforms, at a clock rate of 30.72 MHz once the phase argument θ is sent to the module at that clock rate.

Figure 8 depicts the proposed architecture for the phase to value mapper module, which is in charge of translating values from phase space to time domain. The sine and cosine index values are employed as address values to access the correct positions of the LUT memory. It is the LUT memory that executes the translation from phase space to time domain. The LUT memory stores only 1/4, i.e. $N/4$, samples of the cosine waveform signal employed as the basis waveform signal. The *sel* signal selects whether the sine or cosine index value is employed to access the LUT memory. As it is shown in **Figure 8**, the cosine index value is employed as the address value when the *sel* signal is at low level and the sine index value is used when it is at a high level.

Both sine and cosine index values remain constant for a cycle of the clock rate of 30.72 MHz. Since the LUT memory works at a clock rate of 61.44 MHz and both index values are present at the same time at its input, it is possible to read two samples inside one period of the clock rate of 30.72 MHz. Through the creation of a delayed data path with the use of a register at the output of the LUT memory, it is possible to redirect the I and Q sample components to two distinct data paths and therefore generate the complex exponential sequence that is necessary to translate the received PRACH preamble sequence into baseband. At this stage, the resulting quadrature sequence signal is fully synchronised to the 61.44 MHz clock rate; however, each quadrature pair of values lasts for one period of the 30.72 MHz clock rate. This is explained by the two registers with the chip enable signal inputs set by the *sel* signal that is located at the output of the multiplexers responsible for changing the signal of the quadrature components.

In order to convert the quadrature sample values to their original quadrants, the sine and cosine signals created by the angle mapper module are applied to their respective data paths. When due, the change of signal is easily executed through the application of the complement of two operations to the sample value. Finally, it is necessary to change the clock domain of the complex exponential signal sequence since the ADC module works at a data rate of 30.72×10^6 samples per second. Even though its samples last for the correct period, they are not synchronised to the 30.72 MHz clock rate. The simplest way to execute the clock domain crossing is to use two different registers at the desired clock rate. As we work with complex signals, we employ a pair of dual registers for each one of the quadrature component values. At this stage, the resulting complex exponential sequence signal is totally ready to be multiplied by the received PRACH preamble signal.

4.3 Complex multiplier unit

Figure 9 shows the proposed architecture for the complex multiplier module. A complex multiplier is necessary to multiply the samples coming from both NCO and ADC modules and perform the required frequency shift in time domain, once samples coming from these modules are complex. The complex multiplier module, which is also known as mixer, executes the multiplication of the ADC samples, i.e. the received PRACH sequence signal, by the complex exponential signal created by

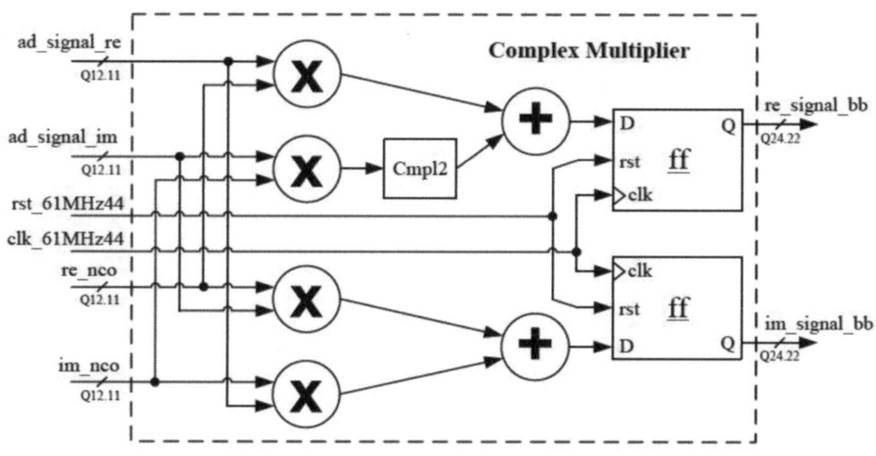

Figure 9.
Architecture of the complex multiplier unit.

the customised NCO module. The multiplication of these two complex values, i.e. $a + jb$ and $c + jd$, results in the complex product defined by Eq. (18):

$$\begin{aligned} real + j * imag &= (a + j * b) * (c + j * d) \\ &= (ac - bd) + j * (ad + bc). \end{aligned} \tag{18}$$

As can be noticed by analysing Eq. (18), the complex multiplication operation needs two additions and four multiplications since a subtraction operation is considered as being an addition in complement of two. The complex multiplier modules works at the clock rate of 30.72 MHz since it must always obey the data rate determined by the ADC module.

5. Implementation and simulation results

In order to assess the efficiency of the customised NCO and time-domain frequency shifter units proposed in this paper, some simulations were carried out. The proposed time-domain frequency shifter architecture was developed in VHSIC hardware description language (VHDL), and a corresponding bit-accurate Matlab model, referred here as Golden Model (GM), was developed for verification. The full design was targeted to a Xilinx Virtex-6 xc6vlx240t FPGA. The results presented next are split into parts: the first one provides the simulation results for the customised NCO architecture implementation, and the second part presents the results regarding the implementation of the time-domain frequency shifter architecture.

5.1 Customised numerically controlled oscillator

This section presents results regarding the customised NCO implementation. The first simulation result, shown in **Figure 10**, compares floating-point precision Matlab-generated complex exponential sequences with fixed-point precision sequences generated by the device under test (DUT) along all possible discrete

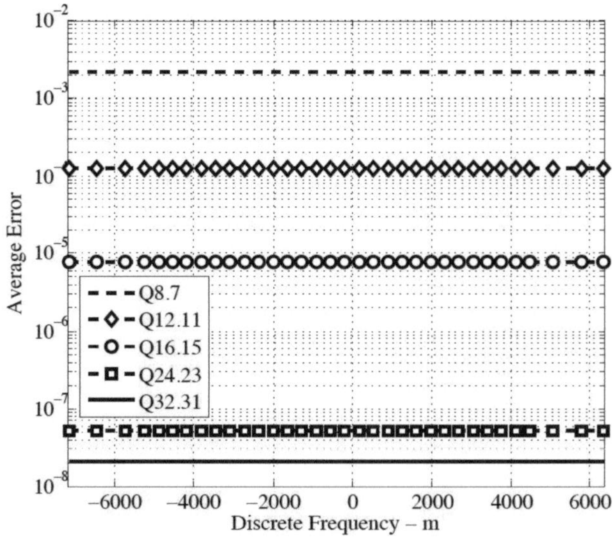

Figure 10.
Average error between GM and DUT implementations of the customised NCO.

frequency shifts, m, and for some Q-formats. PRACH format 0 preambles were considered for this and all other simulation results. **Figure 11** presents the SFDR variation of the implemented NCO unit, i.e. the DUT, along all possible discrete frequency shifts, m, and for some Q-formats. SFDR is the power ratio between the fundamental signal and the strongest spurious signal, i.e. the most prominent harmonic, present at the output of the customised NCO. By analysing this result, it is noticeable that the SFDR attained by the DUT is almost the same for formats Q24.23 and Q32.31. The SFDR values achieved by the DUT for formats Q24.23 and Q32.31 are 153.58 and 154.2 dB, respectively. These high SFDR values are due to the fact that the phase increment bits are not truncated and all output frequencies, f_{out}, are integer multiples of the system clock frequency, f_{clk}, which is the frequency used to sample the basis waveform, therefore eliminating the spectral artefacts resultant from phase jitter [15].

As an illustrative example of the performance presented by the customised NCO, **Figure 12** shows its power spectrum for some fixed-point formats with SFDR indication for frequency shift, m, equal to 7187, i.e. 8,983,750 Hz. These results clearly show the cleanliness achieved by the proposed customised NCO even for format Q8.7. The noise floor for format Q24.23 is so small that it is imperceptible.

Figure 13 depicts the SNR variation of the customised NCO along all possible discrete frequency shifts, m, and for some Q-formats. The SNR results are obtained by the ratio between the signal average power and the noise average power.

5.2 Time-domain frequency shifter

This section presents results regarding the implementation of the time-domain frequency shifter architecture. From this point on, we refer to the architecture implementation as the DUT. This time a Matlab floating-point model (GM) of the whole time-domain frequency shifter architecture is used to assess the performance of the circuit.

Table 4 presents information regarding the resource usage of the proposed architecture. It sums up the key results obtained after the implementation of the proposed frequency shifter architecture on a given FPGA chip. The number of

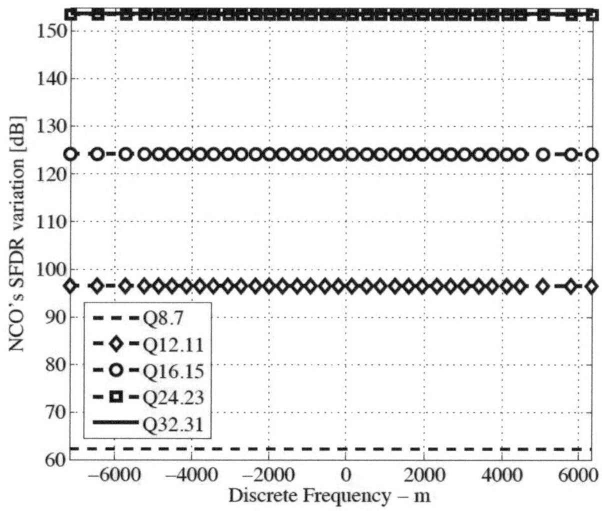

Figure 11.
SFDR variation of the customised NCO.

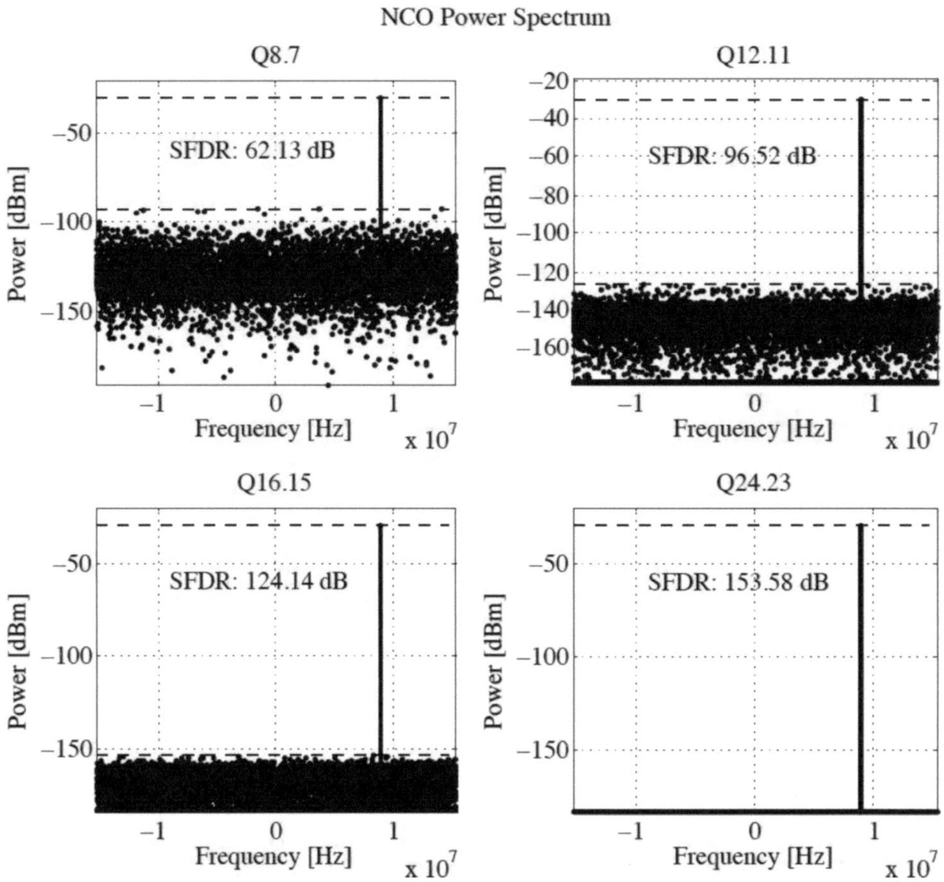

Figure 12.
Customised NCO power spectrum.

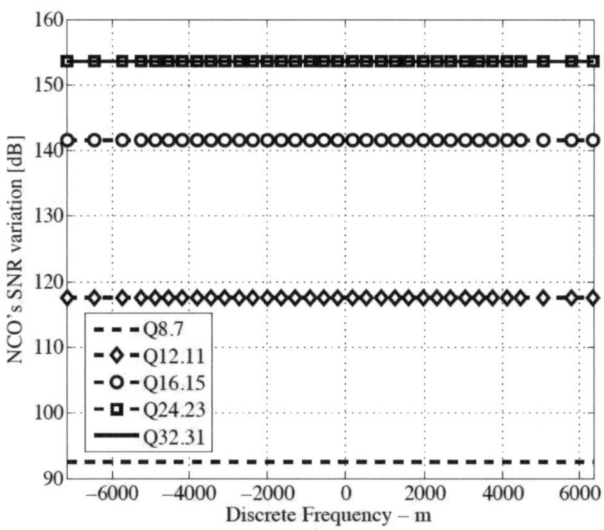

Figure 13.
SNR variation of the customised NCO.

occupied slices, registers, memory resources, LUTs and digital signal processor (DSP) resource blocks is shown in the table. The maximum achievable working frequency that can be reached by the module is equal to 239.981 MHz.

FPGA model number	XC6VLX240T-1ff1156 (Virtex-6)
Amount of slice registers	170 out of 301440 0%
Amount of slice LUTs	215 out of 150720 0%
Amount of occupied slices	84 out of 37,680 0%
Amount of RAMB36E1/FIFO36E1s	3 out of 416 0%
Amount of DSP48E1s	4 out of 768 0%
Maximum achievable frequency	239.981 MHz

Table 4.
Resource usage.

After observing the results presented in **Table 4**, we realise that three block RAM (BRAM) memory resources are employed instead of the two mentioned before. This is explained due to the fact that the synthesis tool maps all the contents of the LUT memory into three BRAM resources since the number of bits employed to address the LUT memory is equal to 13, and therefore, $\text{ceil}((2^{13} * 12)/36K) = 3$ instead of the two BRAMs mentioned earlier. Noticed that each address position of the BRAM resources saves a 12 bit value that is sampled from the basis cosine waveform. The four Xilinx DSP48 resources that are instantiated are used in the complex multiplier module to implement the multiplication operation defined in Eq. (18).

It is important to mention that in Virtex-6 family of FPGAs, one slice consists of eight flip-flops and four LUTs. Block RAMs and FIFO resources are embedded resources of the 36 bit memory resources. A DSP48 resource is an embedded processing unit that corresponds to one multiplier with two 18 bit inputs and one accumulator of 48 bits.

The reuse of DSP48 units is an important and feasible approach that is able to optimise the FPGA area utilisation at the expense of a higher clock rate operation and additional usage of control logical resources. For instance, the four fully parallel multiplications in the complex multiplier module could be serialised, which would save three out of the four DSP48 already being used.

After analysing **Table 4**, it is possible to notice that the proposed frequency shifter architecture uses less than 1% of all available Virtex-6 logical resources. Given that the actual implementation of the proposed architecture on FPGA presents a very low occupancy rate, the utilisation of low-cost FPGA models is possible. Therefore, there are two important points that must be taken into consideration when selecting a low-cost FPGA model: (i) the maximum achievable frequency operation, once low-cost FPGA models tend to present worse timing characteristics, and (ii) the number of used slices might increase in the case of families earlier than the Virtex-6 family, since other families may employ LUT memories with 4 bits instead of LUT memories with 6 bits per slice.

In **Figure 14**, the average error between the frequency shifted preambles generated by the GM and DUT is presented. In order to generate a representative result, the average error for a given RACH preamble is averaged over all possible offsets applied to that RACH preamble. In other words, the figure shows the average error over all possible offsets that can be applied to a given RACH preamble. Moreover, the figure presents the average error for several Q-formats when the PRACH bandwidth parameter, N_{RB}^{UL}, is made equal to 25 and 50 RBs, respectively. These bandwidth parameters correspond to bandwidths of 5 and 10 MHz, respectively. Additionally, the PRACH offset parameter, n_{PRB}^{RA}, is set to all possible values.

Figure 15 shows an exploded view of the results presented in **Figure 14**. Each subplot, representing the error for a specific Q-format, depicts the average of the

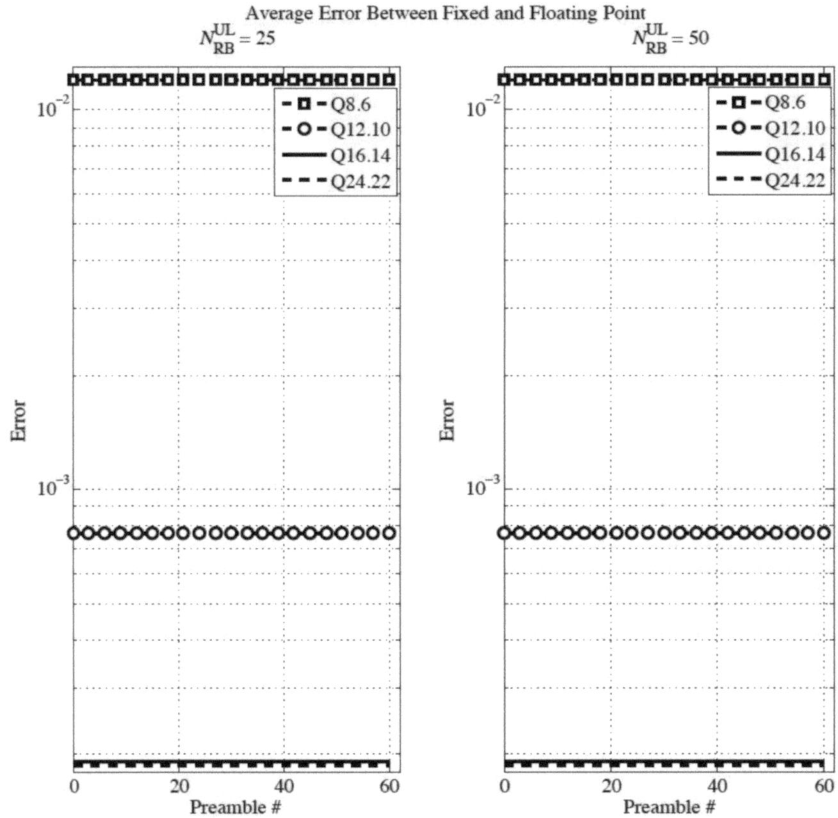

Figure 14.
Average error between GM and DUT.

average error over all possible offsets applied to a given preamble for 25 and 50 RB bandwidths. It is clearly seen that the error has a very small variation, almost constant, along the preambles.

Next we present results regarding the use of the time-domain frequency shifter architecture proposed in this work in the context of the PRACH receiver at eNodeB PHY side. The PRACH receiver architecture adopted in this work is shown in **Figure 2**, and a bit-accurate Matlab model was developed for its verification.

At the receiver side, the eNodeB attempts to detect a transmitted preamble by first extracting the PRACH signal from a received OFDM signal. The extraction involves applying downconversion, analog to digital conversion, CP removal, frequency shift, demapping and decimation to the received PRCAH signal. Next the receiver performs a matched filtering across the pool of preambles allocated to the eNodeB. Matched filtering is performed as a circular cross-correlation between the extracted PRACH signal and each of the known preambles dedicated to the eNodeB.

Figure 2 depicts the preamble detection module, which is the last block in the PRACH receiver processing flow. This module is responsible for detecting the transmission of RACH preambles at the PHY layer. This module employs the detection algorithm proposed in a previous work by the authors of [12]. All samples being received from the IFFT module have their squared modulus computed, then producing what is called as the power delay profile (PDP) samples. This module uses the PDP samples to (i) estimate a noise power value, which is performed by identifying PDP samples that can be regarded as containing only the presence of noise, and (ii) compute a RACH detection threshold. The RACH detection threshold

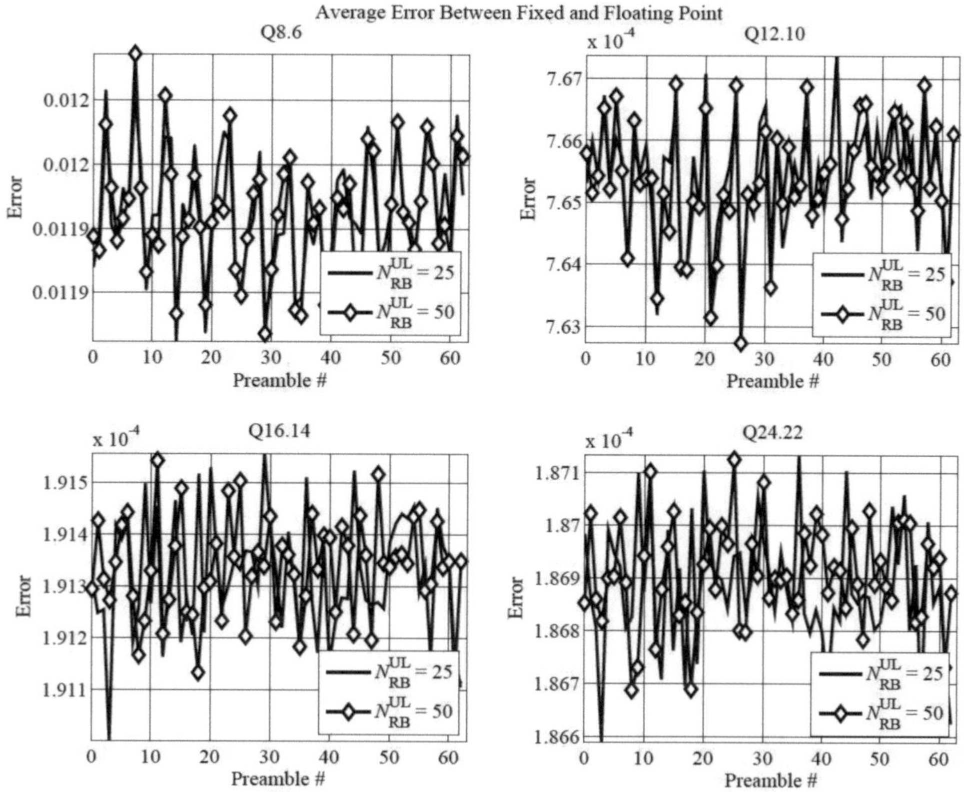

Figure 15.
Exploded view of the average error between GM and DUT.

is calculated based on the noise power estimate that minimises the probability of false alarms. Based on the RACH detection threshold, the PRACH receiver is able to decide whether a RACH preamble is present or not. The RACH detection module reports back to the MAC layer the timing offset estimates and IDs of all detected RACH preambles in a given reception interval. Interested readers are referred to [4, 16] for further details on the PARCH receiver architecture and RACH detection algorithm, respectively.

The bit-accurate PRACH receiver model adopted in this work includes the proposed time-domain frequency shift algorithm. In order to assess the performance of the proposed architecture, format 0 RACH preambles with $N_{CS} = 13$ and corrupted with additive white Gaussian noise (AWGN) were sent to the PRACH receiver model. When N_{CS} is equal to 13, all the 64 RACH preambles that are allocated to a given cell can be created out of a single root ZC sequence.

Through the execution of only one circular cross-correlation operation in the frequency domain between the noisy RACH preambles and the corresponding local root ZC sequence, the PRACH receiver is able to detect random access attempts by several UE devices [4]. The RACH detection process follows the algorithm presented in [12]. For the next results, the bit width of the output data path of the proposed architecture was set to 8, 12, 16 and 24 bits, resulting in the fixed-point representations Q8.6, Q12.10, Q16.14 and Q24.22, respectively.

Figures 16 and **17** present the complementary results of preamble detection when the time-domain frequency shifter is employed along with the preamble detection algorithm proposed in [12]. They depict comparisons between floating-

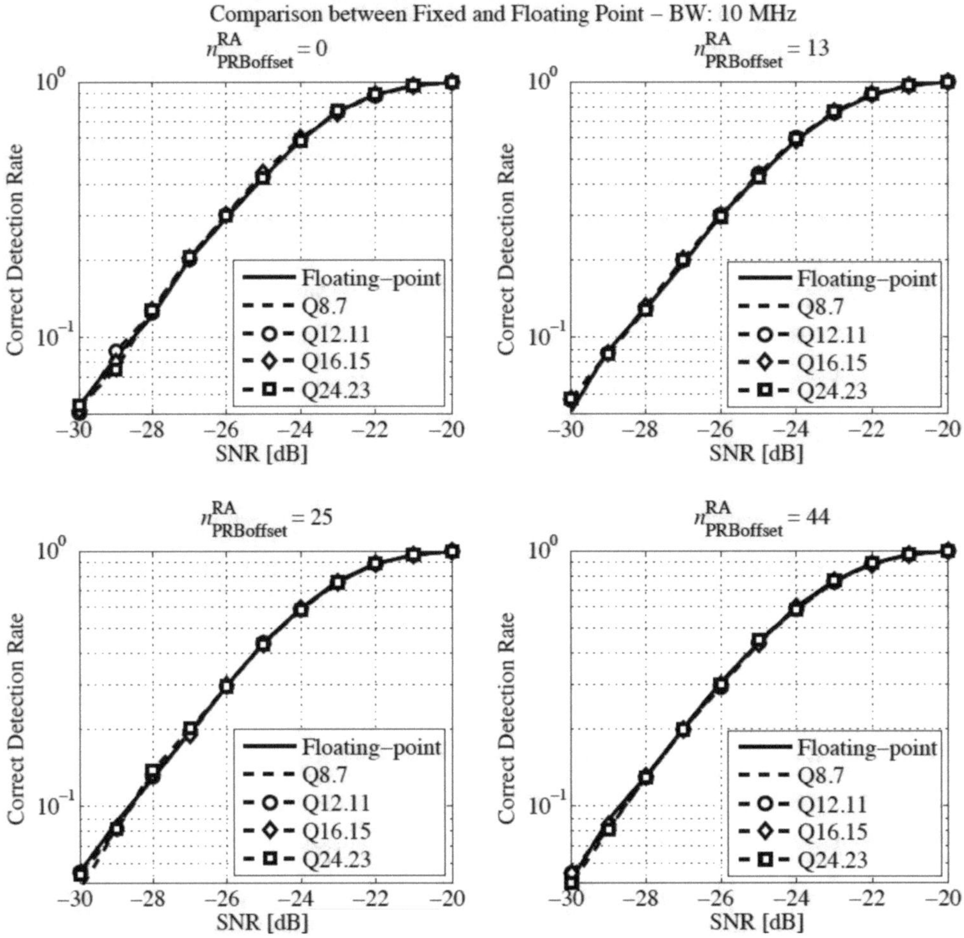

Figure 16.
Comparison of the correct detection rate between the bit-accurate and the floating-point model.

point and fixed-point detection results when bandwidth parameter, N_{RB}^{UL}, is set to 50 RBs (10 MHz).

Each subplot in **Figure 16** presents the comparison of the achieved correct detection rates versus signal-to-noise ratio (SNR) in dB for a given offset, n_{PRB}^{RA}. **Figure 17** presents the comparison of the achieved error detection rates versus SNR for a given offset. For both plots the probability of false alarm (P_{fa}) is made equal to 0.1%. The plots demonstrate the high accuracy of the proposed algorithm and corresponding architecture in translating the received PRACH preambles to baseband.

An important requirement for the PRACH receiver is that it must be capable of serving a huge number of UE devices per cell maintaining a reasonable detection probability and providing them with quasi-instantaneous access to the radio resources, while keeping the false alarm rate to low levels. The probability of a correct detection of the RACH preambles at the receiver side ought to be greater than or equal to 99% at an SNR of −8.0 dB, as defined in Section 8.3.4.1 of [17].

By analysing **Figure 16**, it is possible to see that the proposed algorithm achieves a probability of correct detection greater than 99% at a SNR of −21 dB, clearly outperforming [17] in 13 dB.

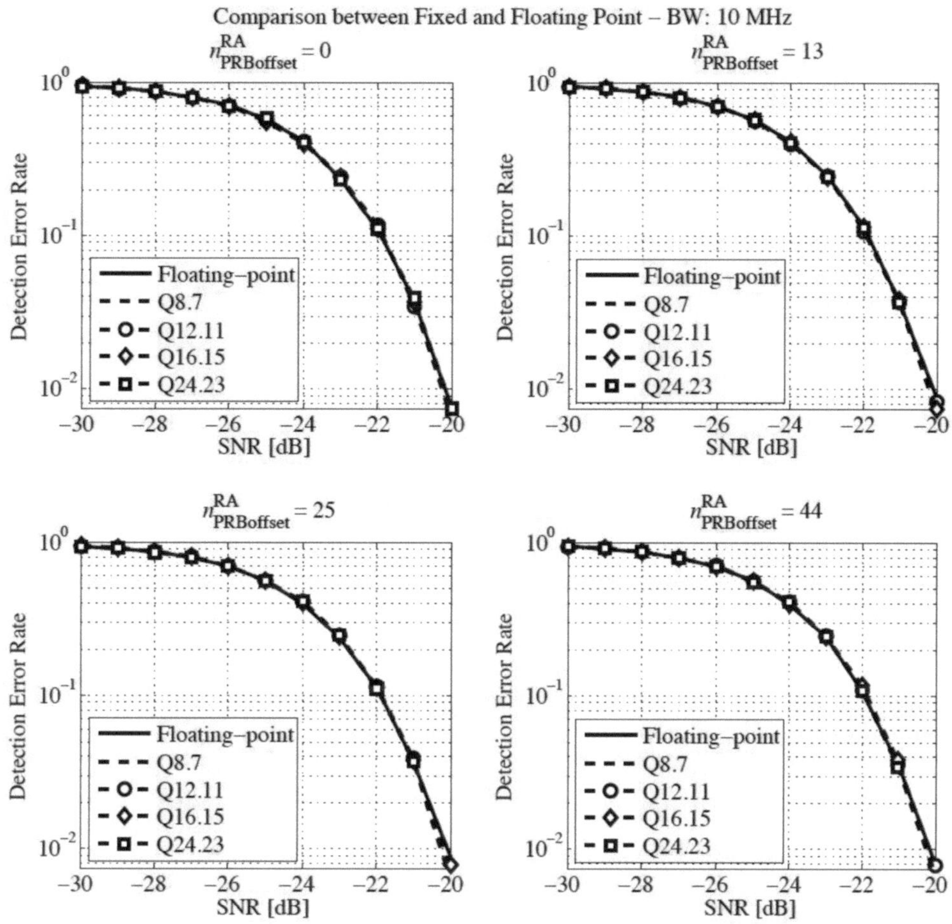

Figure 17.
Comparison of the error detection rate between the bit-accurate and the floating-point model.

6. Conclusions

A hardware-efficient algorithm and architecture for translating PRACH preambles into baseband featuring high-accuracy and low-complexity characteristics has been presented. This paper is an extension of a previous work where we only introduced some superficial aspects of the proposed hardware architecture. In this paper we present theoretical derivations showing how to arrive at the equations used to design and implement the proposed architecture. We provide the pseudocode for the proposed algorithm and discuss the advantage of the proposed architecture in terms of memory utilisation when comparing our proposed solution with an approach where the full period of the complex exponential is stored in FPGA memory. The proposed architecture is optimised to shrink the use of BRAMs, multipliers and logical resources. The low resource utilisation exhibited by the proposed architecture demonstrates its feasibility to be employed as part of large physical layer designs or to be used in FPGAs with small amount of logical elements.

The corresponding hardware architecture has been developed and employed in the PRACH receiver. Implementation and simulation results have demonstrated the efficiency, accuracy and low complexity of the proposed algorithm and architecture. Finally, this paper provides detailed information on the architectural design that was tested on an FPGA device for real-time LTE applications.

Author details

Felipe A.P. de Figueiredo[1,2*] and Fabbryccio A.C.M. Cardoso[3]

1 Instituto Nacional de Telecomunicações—INATEL, Santa Rita do Sapucaí, MG, Brazil

2 Department of Information Technology, IDLab, Ghent University—imec, Ghent, Belgium

3 CPqD—Research and Development Center on Telecommunication, Campinas, SP, Brazil

*Address all correspondence to: felipe.figueiredo@inatel.br

References

[1] Zarrinkoub H. Overview of the LTE Physical Layer. Hoboken, New Jersey, USA: John Wiley & Sons; 2014

[2] Kanchi S, Sandilya S, Bhosale D, Pitkar A, Gondhalekar M. Overview of LTE-A technology. In: IEEE Global High Tech Congress on Electronics (GHTCE). 2014

[3] Rumney M. LTE and the Evolution to 4G Wireless: Design and Measurement Challenges. Hoboken, New Jersey, USA: Wiley; 2013

[4] Sesia S, Toufik I, Baker M. LTE—The UMTS Long Term Evolution: From Theory to Practice. Hoboken, New Jersey, USA: John Wiley & Sons; 2011

[5] de Figueiredo FAP, Mathilde FS, Cardoso FACM, Vilela RM, Miranda JP. Efficient FPGA-based implementation of a CAZAC sequence generator for 3GPP LTE. In: IEEE International Conference on Re-ConFigurable Computing and FPGAs (ReConFig 14). 2014

[6] de Andrade TPC, Astudillo CA, Sekijima LR, da Fonseca NLS. The random access procedure in long term evolution networks for the internet of things. IEEE Communications Magazine. 2017;**55**(3):124-131

[7] 3GPP TS 36.211. Physical Channels and Modulation (Release 10). 2009

[8] Figueiredo FAP, Mathilde FS, Figueiredo FL, Cardoso FACM. An FPGA-based time-domain frequency shifter with application to LTE and LTE-A systems. In: IEEE Latin American Symposium on Circuits & Systems (LASCAS). 2015

[9] Frank RL, Zadoff SA, Heimiller R. Phase shift pulse codes with good periodic correlation properties. IRE Transactions on Information Theory. 1961;7:254-257

[10] Chu DC. Polyphase codes with good periodic correlation properties. IEEE Transactions on Information Theory. 1972;**18**(4):531-532

[11] Yang X, Fapojuwo AO. Enhanced preamble detection for PRACH in LTE. In: IEEE Wireless Communications and Networking Conference (WCNC). 2013

[12] de Figueiredo FAP et al. Multi-stage based cross-correlation peak detection for LTE random access preambles. Revista Telecomunicações. 2013;**15**:1-7

[13] Ranabhatt NA, Agarwal S, Bhattar RK, Gandhi PP. Design and implementation of numerical controlled oscillator on FPGA. In: International Conference on Wireless and Optical Communications Networks (WOCN). 2013

[14] Kadam S, Sasidaran D, Awawdeh A, Johnson L, Soderstrand M. Comparison of various numerically controlled oscillators. In: Midwest Symposium on Circuits and Systems (MWSCAS). 2002

[15] Xilinx. DS246—DDS Logic Core Product Specification—v5. 2005

[16] de Figueiredo FAP, Cardoso FACM, Lenzi KG, Bianco Filho JA, Figueiredo FL. A modified ca-cfar method for lte random access detection. In: 7th International Conference on Signal Processing and Communication Systems (ICSPCS). 2013. pp. 1-6

[17] 3GPP TS 36.104. Base Station (BS) radio transmission and reception (Release 10). 2007

Chapter 5

VLSI Implementation of Medical Image Fusion Using DWT-PCA Algorithms

Surya Prasada Rao Borra, Rajesh K. Panakala and Pullakura Rajesh Kumar

Abstract

Nowadays, the usage of DIP is more important in the medical field to identify the activities of the patients related to various diseases. Magnetic Resonance Imaging (MRI) and Computer Tomography (CT) scan images are used to perform the fusion process. In brain medical image, MRI scan is used to show the brain structural information without functional data. But, CT scan image is included the functional data with brain activity. To improve the low dose CT scan, hybrid algorithm is introduced in this paper which is implemented in FPGA. The main objective of this work is to optimize performances of the hardware. This work is implemented in FPGA. The combination of Discrete Wavelet Transform (DWT) and Principle Component Analysis (PCA) is known as hybrid algorithm. The Maximum Selection Rule (MSR) is used to select the high frequency component from DWT. These three algorithms have RTL architecture which is implemented by Verilog code. Application Specified Integrated Chips (ASIC) and Field Programmable Gate Array (FPGA) performances analyzed for the different methods. In 180 nm technology, DWT-PCA-IF architecture achieved 5.145 mm^2 area, 298.25 mW power, and 124 ms delay. From the fused medical image, mean, Standard Deviation (SD), entropy, and Mutual Information (MI) performances are evaluated for DWT-PCA method.

Keywords: application specified integrated chips, discrete wavelet transform, field programmable gate array, principle component analysis, maximum selection rule

1. Introduction

In recent years, Image Fusion (IF) importance has increased rapidly. The process of combining two or more images into one image is called as IF. Through this, all kinds of information are possible to take from the different images [1]. Based on the image stage, the fusion has been classified into two types, those are transform domain and spatial domain fusion [2]. IF is used in so many applications like medical, automated industry, engineering field, military, etc. [3]. Among all those fields, medical field application is more important in IF which helps to identify the human problems [4]. In medical, two major models like MRI and CT scan helps to analyze the normal and abnormal tissue and internal structure of human body because both MRI and CT contain some different information of the human brain [5]. MRI scan is used for soft tissue which detects the skull problems as well as CT scan is used for hard tissue to identify the bone structure [6]. Earlier many techniques used in IF like pixel level based, decision level, and feature level based [7]. Many of the existing algorithm has been used for IF process such as Electrical Capacitance Tomography (ECT) algorithm [8], Non-Subsample Contour let Transform (NSCT) [9], sparse representation and decision [10], Curvelet transform [11], hybrid Entropy concept [12], hybrid Dual tree complex wavelet transform [13], and hybrid IF and image registration [14]. The main problem with these methods is information loss. To check the hardware utilization and improve the efficiency, the IF has been implemented in FPGA. The way of implementation is also different in FPGA. In FPGA, DWT [15], multi model method [16], and configurable pixel level [17] methods have been implemented for IF process. The hardware utilization of these methods is high. To overcome these problems, hybrid algorithm with the maximum selection rule is implemented in this paper. From the DWT, high frequency component signal only processes the MSR and output of this is given to the Inverse DWT. The combination of DWT and PCA is named as hybrid algorithm. The PCA output gives the IF output. These methods implemented in FPGA architecture to improve the efficiency of the IF. At last, FPGA and ASIC performances improved in proposed method compared to conventional methods. Mean, Standard Deviation (SD), Entropy, and Mutual Information (MI) performances also calculated for all the algorithms. The rest of the paper is organized as follows: Section-2 elaborates the literature survey, Section-3 describes the proposed method, Section-4 discusses the experimental results, and Conclusion is summarized in section-5.

2. Literature review

Mishra et al. [18] presented Modified Frei-chen based image fusion method. This method was utilized in Structural Similarity (SS), and contrast in Night Vision (NV) based two-scale decomposition. This method achieved 48%, 15%, and 100% of improvements in total edge transfer, SS, and NV. This architecture was implemented in the Xilinx tool which consumes 4% of resources. This proposed method was analyzed in synopsis tool with 90 nm CMOS technology. This algorithm provides less accuracy and less fusion efficiency.

Bavirisetti and Dhulli [19] proposed two scale image fusion using saliency detection. This method was used for Saliency extraction process, which can highlight the significant information. This works gave better results compared to multiscale fusion technique. This method failed to process the medical images perfectly.

Pemmaraju et al. [20] presented wavelet based image fusion using FPGA. This proposed method was implemented in Xilinx EDK 10.1 using Spartan 3E. This FPGA contains combinational blocks which are flexible for high speed application. This architecture contains memory, flip flops, and LUT. This proposed method was applied to multi focus image fusion. DWT does not provide stationary outputs and low frequency component has less efficiency.

Yang et al. [21] proposed multi model based image fusion based on fuzzy logic. With the help of type 2 fuzzy, NSCT was analyzed using pre-registered source image for getting low and high bands. Low frequency bands are used by local energy algorithm. The proposed fused image was taken with the help of inverse

NSCT with all sub bands. The accuracy, contrast, and versatility was also evaluated. The main drawback of this method is low spatial resolution.

Bhaskar and Munde [22] proposed image fusion using Non-Subsampled Shearlet Transform (NST) in FPGA implementation. Input image was separated into individual image co-efficient using NST. Different rules were applied to fuse the high and low bands. With the help of inverse NST, the fused image was taken. This proposed method was implemented in Xilinx system generator and MATLAB. The power value was reduced in proposed method. But, the hardware utilization of this proposed method is high.

Agarwal and Bedi [23] presented hybrid image fusion for medical diagnosis. In this paper, wavelet and Curvelet transforms were used to perform the IF. The segmented blocks were fused into sub bands using Curvelet transform. The resolution of the fused image is too less which affects quality of the image.

Sanjay et al. [24] proposed IF based on DWT and type-2 fuzzy logic. In this paper, CT and MRI images were fused with the help of hybrid method. The fused low level bands and high level bands were reconstructed to perform the IDWT. This hybrid algorithm fails to use more logic function and analyses the hardware utilization.

3. DWT-PCA-IF architecture

Image Fusion is one of the important processes for obtaining more information from different images. The overall process of image fusion is shown in **Figure 1**.

- The input CT image is read into MATLAB and the pixel is converted to binary value. These binary values are stored in a text file.

- The same process is applied to MRI images also.

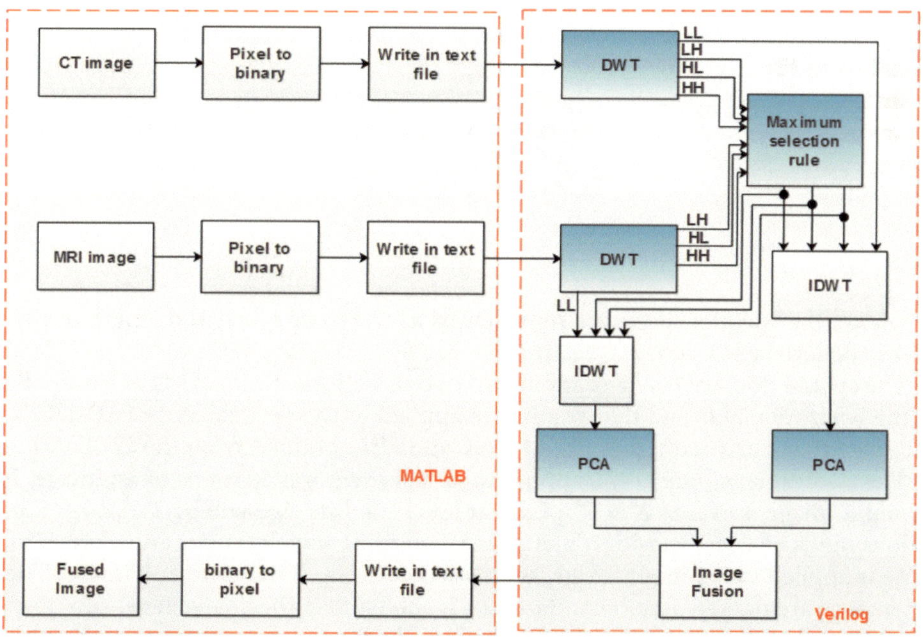

Figure 1.
Block diagram of entire process.

- Both CT and MRI images binary values perform the DWT process which gives four frequency components such as Low High (LH), High Low (HL), High High (HH), and Low Low (LL).
- These frequency components perform MSR. In this operation, high frequency component only required.
- So, HH, HL, and LH frequency components performed MSR operation which gives three results.

These three results are given to the IDWT process along with low frequency component (LL).

- After performing IDWT, both results are given to the PCA component which gives the fused image.
- DWT, MSR, IDWT and PCA are implemented in Verilog and the final output is written in text file.
- With the help of MATLAB, that binary values are converted to pixel which shows the fused image.

3.1 DWT architecture

For analyzing the signal, wavelet converts the time domain to frequency domain. The DWT is implemented using two major blocks namely Filter Bank (FB) and Lifting Scheme (LS). The DWT is a decimated wavelet transform, where the size of the image reduces by half at each scale. It is easy to convert the spatial domain inputs into frequency domain in wavelet transform [25]. High pass and low pass coefficient series are obtained from the input series $y_0, y_1, \dots. y_n$. The high pass and low pass coefficients are represented by using the following two Eqs. (1) and (2).

$$H_i = \sum_{n=0}^{l-1} (2j - n).s_n(z) \tag{1}$$

$$L_i = \sum_{n=0}^{l-1} (2j - n).t_n(z) \tag{2}$$

where, the wavelet filters are represented as $s_n(z)$ and $t_n(z)$, and length of the filter is denoted as l and $j = 0, \dots, [n/2] - 1$.

The spatial domain DWT is applied in two directions. First, 1D-DWT is applied on the horizontal axis and that results are applied to the vertical axis of 1D-DWT. There are four parts named as *LL*, *LH*, *HL* and *HH* obtained from the 2D-DWT.

The two-dimensional DWT applies to all the rows and columns of an image. If the input image is of size $2^k \times 2^k$ pixels at level $L + 1$ its size will be $2^{k/2} \times 2^{k/2}$ The various kinds of decomposition methods are used in wavelets over an image. The DWT is applied to the input image, which is decomposed into four sub image. These sub images are named as sub bands. The *LL* sub band is the coarse level sub image, *HH*, *LH*, and *HL* are the diagonal, vertical and horizontal components of the image respectively. Finally, the input image is decomposed into four major components that is shown in **Figure 2**. A high level 2D-DWT is developed by *LL* frequency and low pass components for multi resolution analysis.

Let assume input image is *Y*.

Here, *Y* is splitting into two different bands such as Y_o and Y_e.

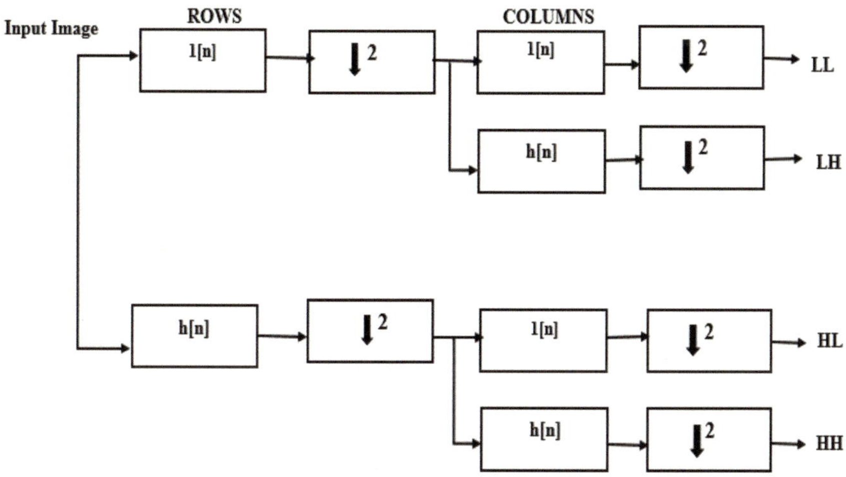

Figure 2.
Discrete wavelet transform.

$$Y_o = [Y(1),\ Y(3),\ Y(5) \ldots Y(2n-1)] \tag{3}$$

$$Y_e = [Y(2),\ Y(4),\ Y(6) \ldots Y(2n)] \tag{4}$$

$$Q_1(n) = Y_o(n) + a(Y_e(n) + Y_e(n+1)) \tag{5}$$

$$V_1(n) = Y_e(n) + b(Q_1(n) + Y_e(n+1)) \tag{6}$$

$dc(n) = L.Q_1(n)$ Here, $Q_1(n)$ is scaled by L.

The 2D-DWT architecture and 1D-DWT are shown in **Figures 3** and **4**. The control signals represent as *clk* and *rst*. The odd input and even input are mentioned as *odd_in* [7:0] and *even_in* [7:0]. These two inputs are given to the line buffer to perform even and odd extraction which outputs are given to the PIPO for capturing the data. From that block, four outputs are generated which is given to the lifting block. After processing the lifting block, the final output is generated as detailed co-efficient *dc_out* [27:0] and significance co-efficient *sc_out* [23:0].

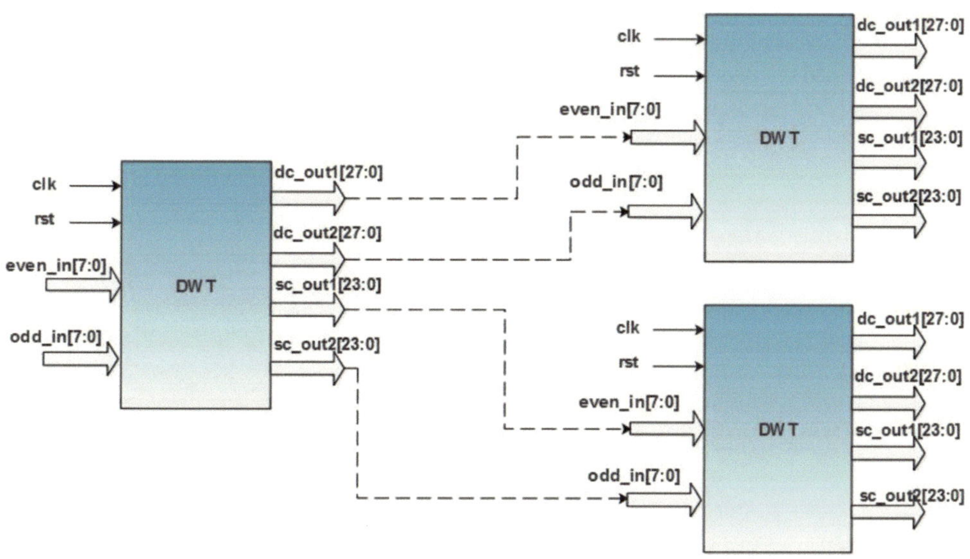

Figure 3.
2D-DWT architecture.

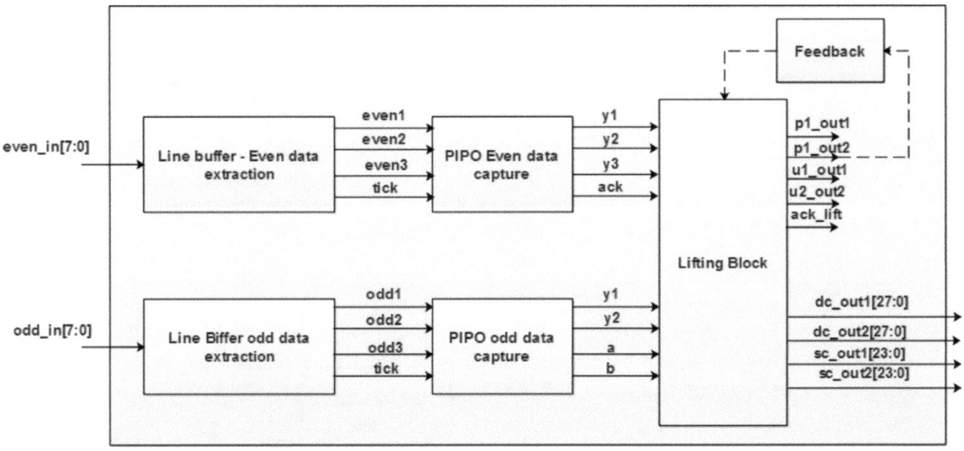

Figure 4.
1D-DWT architecture.

3.2 Maximum selection rule

The MSR diagram is shown in **Figure 5**. This rule is applicable for the high frequency component. So that *HH*, *HL*, *LH* frequency values perform the MSR operation. Both DWT output values are connected to the MUX for choosing maximum value.

These outputs are given to the IDWT for changing the frequency domain into the time domain.

3.3 PCA architecture

The architecture of PCA is shown in **Figure 6** which contains control engine, covariance matrix, MUX, multiplier, adder, and comparator. With the help of detected spike waveform, the covariance matrix is calculated. The covariance matrix is called as PC spike waveform. The MAC address is used for distilling and orthogonalization process to improve the PCA efficiency. Comparator and right shift are used to shift the procedure and level checking. The entire algorithm split into four processing units and the data is stored in register files. Finite State Machine (FSM) is used for scheduling and allocating the resources during the PCA

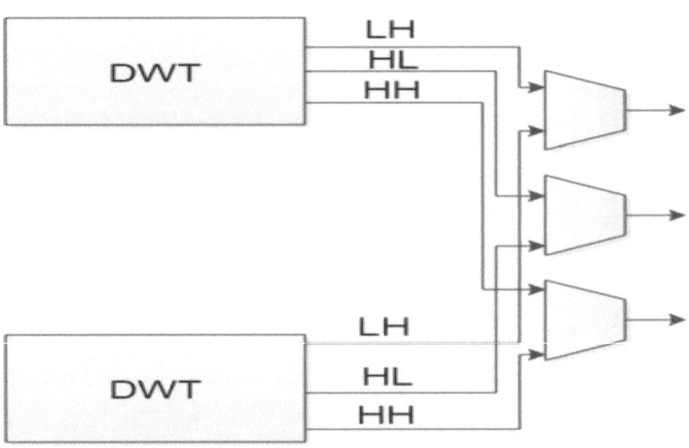

Figure 5.
Maximum selection rule diagram.

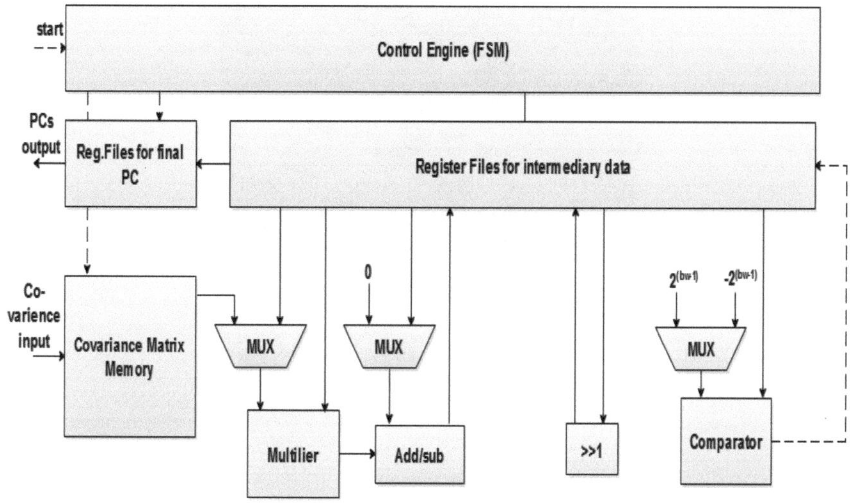

Figure 6.
PCA architecture.

processing. FSM is very effective for controlling the remaining signal [26]. These outputs are helpful to perform the image fusion. The fused architecture binary output is read in MATLAB for showing fused image.

4. Experimental results and discussion

In this section, the experimental simulation results and discussion of the proposed methodology is detailed effectively in terms of performance measure. The performance of the proposed methodology was evaluated by ASIC and FPGA performances.

4.1 Discussion

The input images (CT and MRI) are shown in **Figures 7** and **8**. These images are converted to binary which are shown in **Figures 9** and **10**.

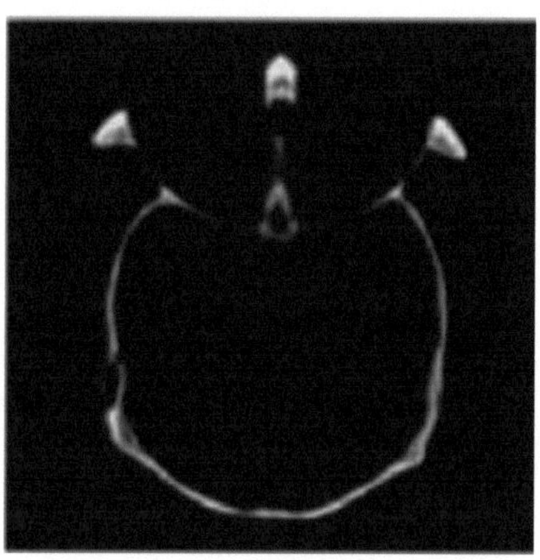

Figure 7.
Input CT image.

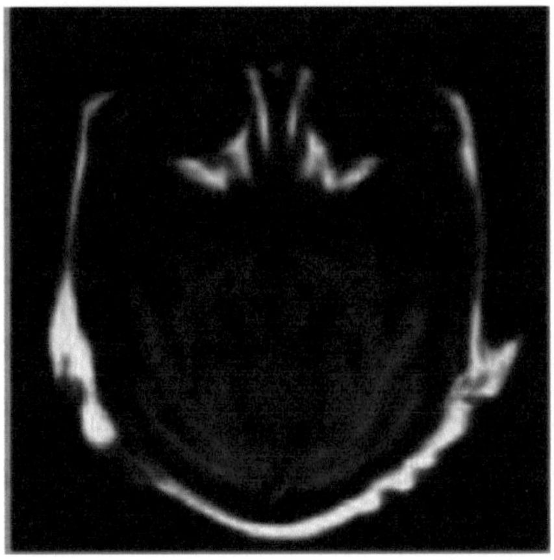

Figure 8.
Input MRI image.

10000011
01000001
11001100
00111011
01000001
10111011
10110101
10110011
00010011
01001110
00011000
01000110
11111001
11111010

Figure 9.
Binary value of CT image.

The ASIC performance of the different methods are tabulated in **Table 1**. In this table, values of ASIC performance of the Existing-I [18], existing-II [20], existing-III [22], and DWT-PCA-IF are compared.

The comparison of ASIC performances is tabulated in **Table 1**. Here, all the methods are implemented and the results are tabulated. All the methods are implemented in the cadence RTL compiler with 180 and 45 nm technology. From this table, it's clear that DWT-PCA-IF provides better performances when compared to previous existing architectures.

Figure 10.
Binary value of MRI image.

Technology	Method	Area (mm^2)	Power (mW)	Delay (ms)
180 nm	Existing-I [18]	8.471	387.1	180
	Existing-II [20]	7.321	345.71	158
	Existing-III [22]	6.214	314.21	143
	DWT-PCA-IF	5.145	298.25	124
45 nm	Existing-I [18]	3.014	198.25	104
	Existing-II [20]	2.987	168.12	101
	Existing-III [22]	2.158	148.687	98
	DWT-PCA-IF	1.982	111.21	91

Table 1.
Comparison of area, power, and delay for different methods.

4.2 Comparative analysis

In this work, three papers have been compared with proposed method. A. Mishra, S. Mahapatra, and S. Banerjee [18], applied modified Frei-chen operator based IF for real time applications. Scalable decomposition was used to perform the fusion operation which was implemented in Virtex 4 FPGA. The overall architecture RTL was too complex to perform the IF algorithm which caused more area. Pemmaraju et al. [20], implemented IF based on DWT using FPGA. This algorithm was implemented in Xilinx EDK 10.1 FPGA Spartan 3E hardware. There is no explanation of RTL architecture, and .ucf file. Due to use of wavelet, the power consumption is too high. Bhaskar and Munde [22] performed IF based on non-subsampled shearlet transform. Xilinx system generator was used to implement this design with MATLAB tool. The fused image affected by more noise and it require more hardware utilization.

The comparison graph of area, power, and delay are shown in **Figures 11–13**. The dark blue bar graph is represented as DWT-PCA-IF architecture. All the ASIC performance is reduced due to the hybrid algorithm.

The FPGA performances are tabulated in **Table 2**. In this table, Virtex 4 and Virtex 5 devices are used to evaluate LUT, flip flop, slices, and frequency. These values are shows that the DWT-PCA-IF architecture achieves better FPGA performance parameters.

The comparison graph of LUT, Flip flop, slices, and frequency are shown in **Figures 14–17**. The hardware utilizations are evaluated from this FPGA performance. The top module and 2D DWT and 1D DWT RTL schematic diagram are shown in **Figures 18** and **19**.

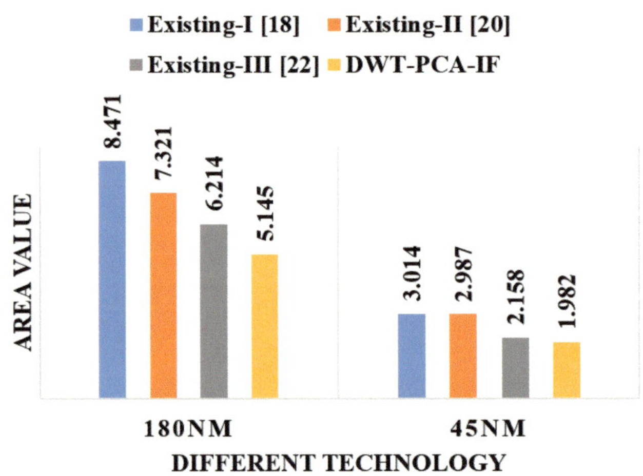

Figure 11.
Comparison of area performance for different methods.

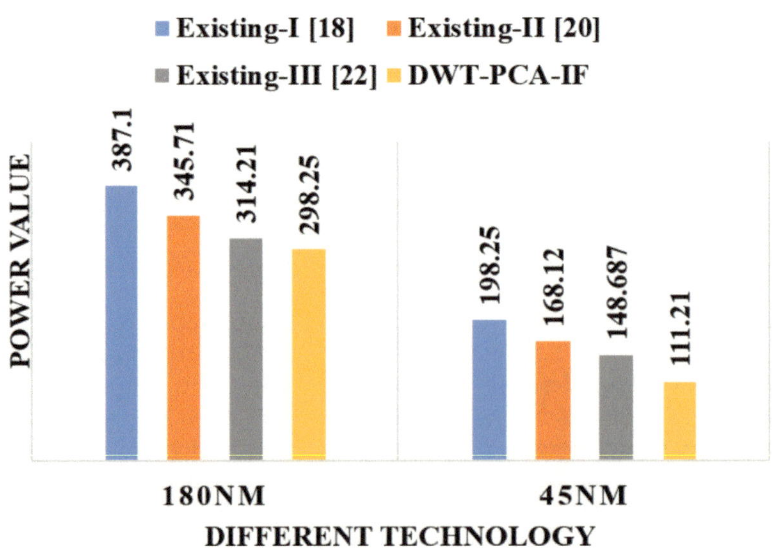

Figure 12.
Comparison of power performance for different methods.

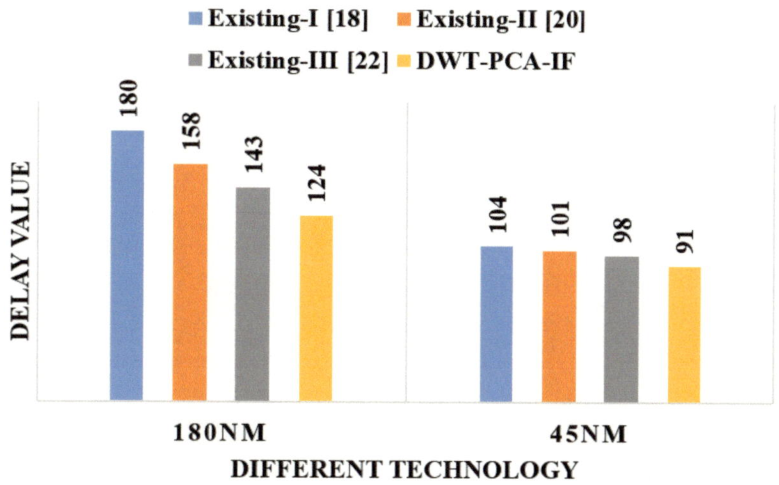

Figure 13.
Comparison of delay performance for different methods.

Devices	Method	LUT	Flip flop	slices	Frequency
Virtex 4	Existing-I [18]	4038	4852	2857	250.3
	Existing-II [20]	4002	4657	2654	289.64
	Existing-III [22]	3541	4214	2011	314.21
	DWT-PCA-IF	3014	3987	1968	355.14
Virtex 5	Existing-I [18]	3104	4125	1964	185.41
	Existing-II [20]	3014	4032	1847	193.21
	Existing-III [22]	2987	3987	1752	255.14
	DWT-PCA-IF	2741	3789	1648	287.96

Table 2.
Comparison of FPGA performances for different methods.

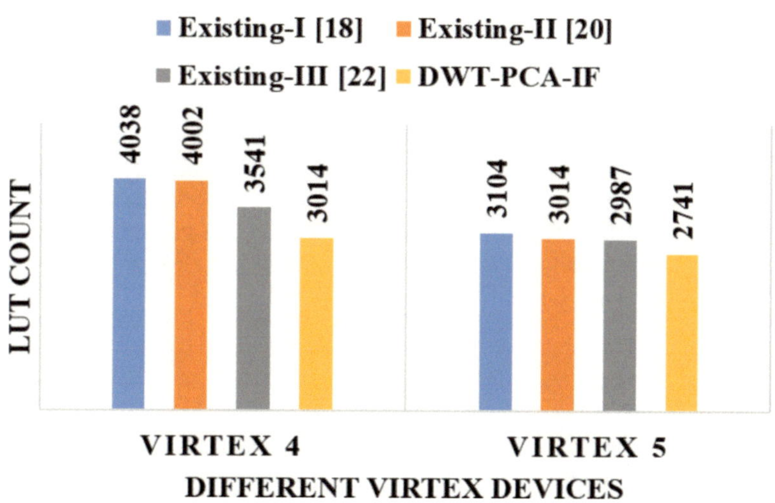

Figure 14.
Comparison of LUT for different methods.

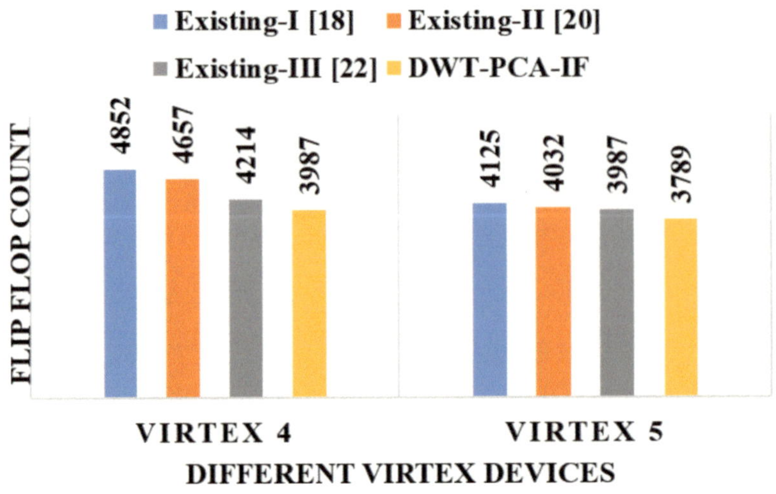

Figure 15.
Comparison of flip flop for different methods.

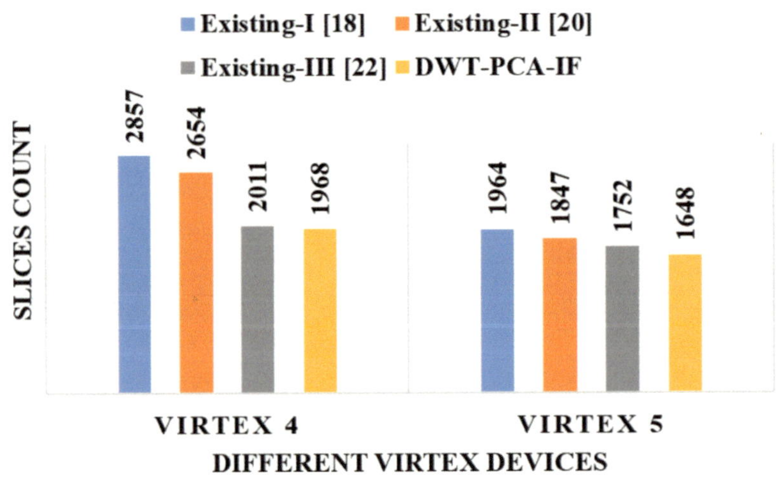

Figure 16.
Comparison of slices for different methods.

The performance evaluation for different methods is given in **Table 3**. Here, some of the performances are evaluated such as Mean, Standard Deviation (SD), Entropy, and Mutual information (MI). This performance evaluated for fused medical image. From this table, it is clears that DWT-PCA gives better performances than existing methods. Finally, the fused image is shown in **Figure 20**. The above RTL schematics are taken from the Xilinx tool.

5. Conclusion

The proposed architecture has been designed effectively in order to reduce the hardware utilization. In this work, DWT-PCA-IF architecture has been designed to perform the image fusion. In this work, medical images like MRI and CT have been used in the fusion process to obtain more information. The hybrid VLSI architecture

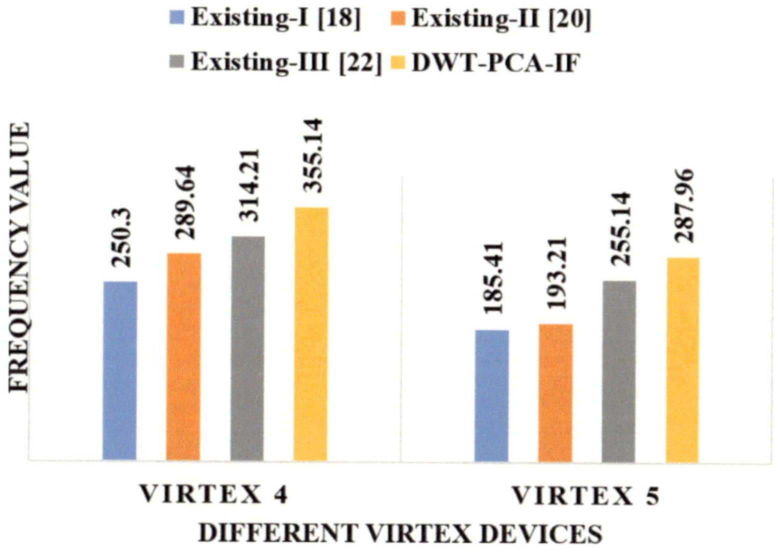

Figure 17.
Comparison of frequency for different methods.

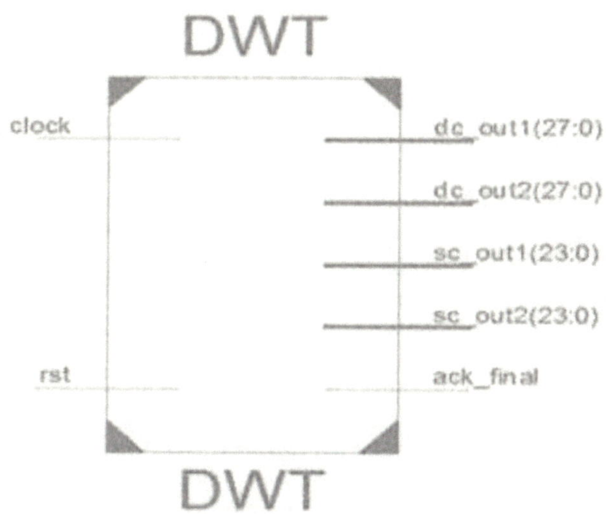

Figure 18.
Top module of DWT architecture.

provided better fused image compared to previous works. The DWT-PCA-IF architecture was implemented using Verilog code. DWT and PCA method were used to reduce the power and area consumption. The ASIC and FPGA performance were analyzed for different architectures. In 180 nm technology, DWT-PCA-IF architecture achieved 5.145 mm^2 area, 298.25 mW power, and 124 ms delay. In Virtex 4, the proposed architecture achieved 3014 LUT, 3987 flip flop, 1968 slices, and 355.14 MHz frequency. From the fused image, 55.658 mean, 53.14 SD, 9.621 entropy, and 3.141 MI value has been evaluated. In the future, different kind of optimization algorithm will be designed to improve the ASIC and FPGA performances.

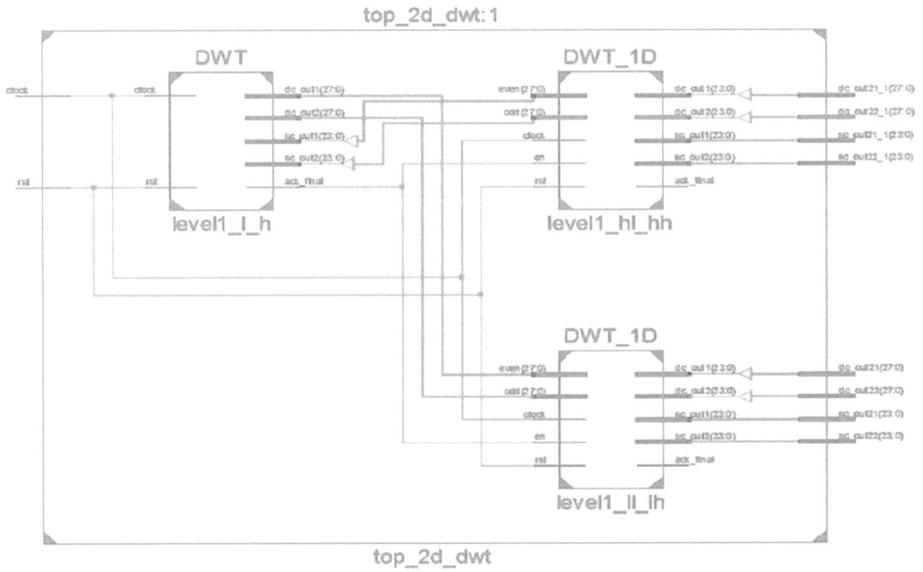

Figure 19.
Internal schematic of 2D-DWT.

Image	Performance	DWT [2]	Haar [3]	Kekre's wavelet [7]	DTCWT [13]	PCA [24]	DWT-PCA
Fused Image	Mean	44.25	32.53	32.41	45.14	53.22	55.65
	SD	40.14	36.07	34.82	51.24	37.44	53.14
	Entropy	8.145	5.97	5.9108	47.21	6.63	9.621
	MI	0.147	0.39	0.5541	2.12	0.2832	3.141

Table 3.
Performance evaluation for different methods.

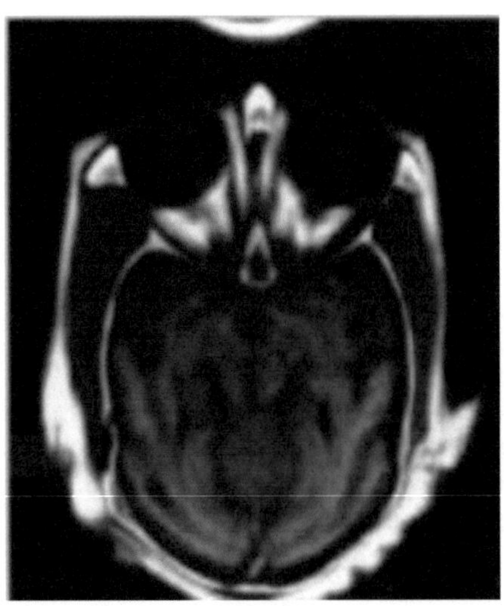

Figure 20.
Fused image.

Acknowledgements

At the outset, I would like to take this as an opportunity to convey my gratitude to Intechopen publishing house and their team to for their consistent support at every step in bringing out this chapter in their book. The process that followed in reviewing this chapter and giving valuable review remarks helped a lot to meet the standards of this book and IntechOpen publishing house has enriched my writing skills. I would like to extend my sincere gratitude to my Ph.D. supervisor Dr. Rajesh K Panakala and Dr.P.Rajesh Kumar for valuable guidance and continuous encouragement in publishing this chapter. I would like to thank my family members for their love and support and the management of our college, PVP Siddhartha Institute of Technology for their constant encouragement to carryout my research work.

Author details

Surya Prasada Rao Borra[1*], Rajesh K. Panakala[1] and Pullakura Rajesh Kumar[2]

1 Prasad V. Potluri Siddhartha Institute of Technology, Kanuru, A.P., India

2 Andhra University College of Engineering, Visakhapatnam, A.P., India

*Address all correspondence to: suryaborra1679@gmail.com

References

[1] Mahajan S, Singh A. A comparative analysis of different image fusion techniques. IPASJ International Journal of Computer Science. 2014;**2**(1):008-015

[2] Hussain DK, Reddy CL, Kumar VA. Implementation of medical image fusion using DWT process on FPGA. International Journal of Computer Applications Technology and Research. 2013;**2**(6):676-679

[3] Phanindra P, Babu JC, Shree VU. VLSI implementation of medical image fusion using Haar transform. International Journal of Scientific and Engineering Research. 2013;**4**(9): 1437-1442

[4] Yang B, Li S. Pixel-level image fusion with simultaneous orthogonal matching pursuit. Information Fusion. 2012;**13**(1): 10-19

[5] Pavithra C, Bhargavi S. Fusion of two images based on wavelet transform. International Journal of Innovative Research in Science, Engineering and Technology. 2013;**2**(5):1814-1819

[6] Jose B, Kumar BS. Design of 2-D DWT VLSI architecture for image processing. International Journal of Engineering Research and Technology. 2014;**3**(4):692-696

[7] B. Kekre H, Sarode T, Dhannawat R. Implementation and comparison of different transform techniques using kekre's wavelet transform for image fusion. International Journal of Computer Applications. 2012;**44**(10): 41-48

[8] Olmos AM, Botella G, Castillo E, Morales DP, Banqueri J, García A. A reconstruction method for electrical capacitance tomography based on image fusion techniques. Digital Signal Processing. 2012;**22**(6):885-893

[9] Bhatnagar G, Wu QMJ, Liu Z. Directive contrast based multimodal medical image fusion in NSCT domain. IEEE Transactions on Multimedia. 2013; **15**(5):1014-1024

[10] Fei Y, Wei G, Zongxi S. Medical image fusion based on feature extraction and sparse representation. International Journal of Biomedical Imaging. 2017; **2017**:1-11

[11] Tank VP, Shah DD, Vyas TV, Chotaliya SB, Manavadaria MS. Image fusion based on wavelet and Curvelet transform. IOSR Journal of VLSI and Signal Processing. 2013;**1**(5):32-36

[12] Sharmila K, Rajkumar S, Vijayarajan V. Hybrid method for multimodality medical image fusion using discrete wavelet transform and entropy concepts with quantitative analysis. In: Proceedings of International Conference on Communications and Signal Processing. 2013. pp. 489-493

[13] Gurjar R. Hybrid image fusion implemented in DTCWT. International Journal of Engineering Technology and Computer Research. 2014;**2**(1):688-692

[14] Bhosle DS, Gorde KS. Image registration and wavelet based hybrid image fusion. IOSR Journal of VLSI and Signal Processing. 2014;**4**(2):1-5

[15] Suraj AA, Francis M, Kavya TS, Nirmal TM. Discrete wavelet transform based image fusion and de-noising in FPGA. Journal of Electrical Systems and Information Technology. 2014;**1**(1): 72-81

[16] Kaur R, Kaur S. An approach for image fusion using PCA and genetic algorithm. International Journal of Computer Applications. 2016;**145**(6): 54-59

[17] Besiris D, Tsagaris V, Fragoulis N, Theoharatos C. An FPGA-based hardware implementation of configurable pixel-level color image fusion. IEEE Transactions on Geoscience and Remote Sensing. 2012; **50**(2):362

[18] Mishra A, Mahapatra S, Banerjee S. Modified Frei-Chen operator-based infrared and visible sensor image fusion for real-time applications. IEEE Sensors Journal. 2017;**17**(14):4639-4646

[19] Bavirisetti DP, Dhuli R. Two-scale image fusion of visible and infrared images using saliency detection. Infrared Physics & Technology. 2016; **76**:52-64

[20] Pemmaraju M, Mashetty SC, Aruva S, Saduvelly M, Edara BB. Implementation of image fusion based on wavelet domain using FPGA. In: Proceedings of International Conference on Trends in Electronics and Informatics. 2017. pp. 500-504

[21] Yang Y, Que Y, Huang S, Lin P. Multimodal sensor medical image fusion based on type-2 fuzzy logic in NSCT domain. IEEE Sensors Journal. 2016; **16**(10):3735-3745

[22] Bhaskar PC, Munde MV. FPGA implementation of non-subsampled Shearlet transform for image fusion. In: Proceedings of International Conference on Computing, Communication, Control and Automation. 2017. pp. 1-6

[23] Agarwal J, Bedi SS. Implementation of hybrid image fusion technique for feature enhancement in medical diagnosis. Human-Centric Computing and Information Sciences. 2015;**5**(1):3

[24] Sanjay AR, Soundrapandiyan R, Karuppiah M, Ganapathy R. CT and MRI image fusion based on discrete wavelet transform and Type-2 fuzzy logic. International Journal of Intelligent Engineering and Systems. 2017;**10**(3): 355-362

[25] Surya PRB, Panakala RK, Kumar PR. Hybrid image fusion algorithm using DWT maximum selection rule and PCA. International Journal of Scientific and Engineering Research;**8**(8):814-820

[26] Surya PRB, Panakala RK, Kumar PR. Qualitative analysis of MRI and enhanced low dose CT scan image fusion. In: Proceedings of International Conference on Advanced Computing and Communication Systems. 2017. pp. 1752-1757